BUILDING REVOLUTIONS

APPLYING THE CIRCULAR ECONOMY TO THE BUILT ENVIRONMENT

RIBA ## **Publishing**

David Cheshire

Kindly supported by

AECOM

Building Revolutions

© AECOM Ltd, 2016

Published by RIBA Publishing, part of RIBA Enterprises Ltd,
The Old Post Office, St Nicholas Street, Newcastle upon Tyne, NE1 1RH

ISBN 978 1 85946 645 2
Stock code 86743

The right of David Cheshire to be identified as the Author of this Work has been asserted in accordance with the Copyright, Designs and Patents Act 1988 sections 77 and 78.

All rights reserved. No part of this publication may be reproduced, stored in a retrieval system, or transmitted, in any form or by any means, electronic, mechanical, photocopying, recording or otherwise, without prior permission of the copyright owner.

British Library Cataloguing-in-Publication Data
A catalogue record for this book is available from the British Library.

Publisher: Steven Cross
Commissioning Editor: Fay Gibbons
Production: Richard Blackburn
Design: Philip Handley
Cover Design: Michèle Woodger
Cover images: James Hager/Robert Harding (dung beetle);
　　　　　　　　iStock.com/LifesizeImages (aerial – fisheye lens)
Typeset: Academic + Technical, Bristol
Printed and bound by Page Bros, Norwich

While every effort has been made to check the accuracy and quality of the information given in this publication, neither the Author nor the Publisher accept any responsibility for the subsequent use of this information, for any errors or omissions that it may contain, or for any misunderstandings arising from it.

www.ribaenterprises.com

About the author

David Cheshire is a Regional Director at AECOM, specialising in sustainability in the built environment. David has over twenty years' experience acting as a sustainability champion on construction projects and has written best practice industry guidance, including the CIBSE's *Sustainability Guide*. David trained as a building surveyor and has an MSc in Energy and the Built Environment. He is a Chartered Environmentalist, a BREEAM Accredited Professional and sits on the SKA Rating technical committee, as well as delivering training on sustainable buildings. He is an advocate of the transition to a circular economy.

Acknowledgements

The idea for this book was inspired by reading *Cradle to Cradle* (by William McDonough and Michael Braungart) and the Ellen MacArthur Foundation's *Towards a Circular Economy* report. The *Cradle to Cradle* book convinced me that design could be used to make buildings '100% good' rather than just 'less bad' and the Ellen MacArthur report made me realise that there is an immediate and compelling argument for implementing a more circular economy.

I'd like to thank Miles Attenborough, David Weight, Andrew Cripps and Ant Wilson from AECOM for their support and encouragement; Richard Hind at UCL for sharing his vast expertise and enthusiasm; and the reviewers, Richard Francis and Charlie Law, for their insightful comments. I'm indebted to the contributors who so generously shared their work, case studies, images, ideas and time to help me create this book; they are all credited in the case studies throughout this book. I'm also grateful to RIBA Publishing's excellent commissioning and editorial team.

Lastly, I'd like to thank my wife, Jolanta, for her support and for giving me the time to write.

Contents

iii	About the author
iv	Acknowledgements
vi	Foreword
1	Introduction

3	01	What is a circular economy?
13	02	Why create a circular economy?
19	03	Built to last?
25	04	Starting at the end
31	05	Circular economy principles for buildings
35	06	Building in layers
41	07	Designing-out waste
53	08	Design for Adaptability
65	09	Design for disassembly and reuse
81	10	Selecting materials and products
101	11	Turning waste into a resource
109	12	Circular business models
117	13	Virtuous circles
127	14	Coming full circle

131	References
135	Index
137	Image Credits

Foreword

The construction industry accounts for approximately 60% of UK materials use and one third of all waste arisings. Buildings are stripped out every few years and often torn down well short of their design life with hardly any products or materials being reclaimed for reuse. This linear model of 'take, make and dispose' is depleting the world's precious resources and is creating mountains of waste with very little scope for reclamation.

In a circular economy, resources are kept in use and their value is retained. This starts at the design stage, where products are designed for disassembly and reuse and new business models incentivise reclaiming, refurbishing or remanufacturing products. In the building industry's current, linear model, topics such as materials selection, waste reduction, resource efficiency, adaptability and design for deconstruction are at best treated as separate issues. At worst they fall short of being considered due to the fragmentation of responsibility within the construction industry, as each discipline blames the next for a lack of holistic thinking or long-term vision.

Applying circular economy thinking to buildings provides an opportunity to draw together all of these seemingly disparate performance aspects into a cohesive whole that creates multiple benefits for people and the environment.

In this book, David Cheshire has created a simple framework and a set of circular economy principles specifically for the built environment. He goes on to explain each of these principles and uses case study examples to show how construction clients, designers and occupants can create a more regenerative built environment.

The case studies show that applying circular economy principles to buildings ensures that they use less resources, can be adapted to different uses and even provide healthier environments for people to live and work in. At the same time, the total cost of ownership can be reduced by engaging with the supply chain and by applying leasing models to shift the cost of upgrade and disposal of equipment back to the manufacturer.

Instead of designing buildings like there is no tomorrow, we need to think about the future of buildings and their users. The profligate attitude we have so far applied to construction resources will not be compatible with the priorities of future generations. It is high time we addressed this, and recognised our stewardship role over the planet's finite resources. Such a transformation of the built environment industry is at the heart of the UK Green Building Council's mission, so I welcome the work David Cheshire has put into articulating a different future for an industry with so much potential for greater efficiencies.

Julie Hirigoyen, CEO, UK Green Building Council

Introduction

A circular economy is an industrial system that is restorative or regenerative by intention and design.
                                          Ellen MacArthur Foundation

Is it the End of the Line for a Linear Economy?

The disposable society is plundering the world of precious, finite resources at an increasing rate.[1] Buildings are stripped out and torn down with astonishing regularity, while new buildings are constructed from components made from hard-won virgin materials. There is little – or no – consideration of the environmental and social impacts of winning and processing construction materials and there is no real thought about the fate of our buildings once they have been built.

The built environment demands around 40% of the world's extracted materials and waste from demolition and construction represents the largest single waste stream in many countries.[2] The raw materials required for our built environment are becoming harder to extract[3] and are putting more strain on the environment as fragile ecosystems are exploited.[4] This is combined with geopolitical forces that cause price volatility and disruptions in the supply of essential raw materials. Meanwhile, demand for resources is predicted to rise as the global middle class is set to double in size by 2030.[5] Global steel demand alone is predicted to rise by 50% by 2025.[6]

Building design is, all too often, tailored to the current users and constructed with little thought of the future. Buildings are also composed of complex components with a bewildering amount of different materials and polymers melded together irretrievably. This lack of regard for the future life of buildings risks leaving a legacy of obsolete architecture with precious resources locked up and away from future generations.

These systemic problems represent a huge challenge for the construction industry and the predicted growth in demand for raw materials is going to put further pressure on a resource-constrained world.

The Birth of a New Economic Model

A new model is emerging, one that moves away from this current 'linear economy', where materials are mined, manufactured, used and thrown away, to a more circular economy where resources are kept in use and their value is retained.

For buildings, this means creating a regenerative built environment that prioritises retention and refurbishment over demolition and rebuilding. It means designing buildings that can be adapted, reconstructed and deconstructed to extend their life and that allow components and materials to be salvaged for reuse or recycling.

New business models allow short-lived elements of the building to be leased instead of purchased, providing occupants with increased flexibility and the ability to procure a service rather than having the burden of ownership. Building collaborative relationships enables manufacturers to invest in product development instead of having to focus on the next sale. Creating a demand for reclaimed or remanufactured components will stimulate the local economy and create new industries, while reducing waste.

A forensic examination of the materials that go into buildings ensures that any potentially harmful substances are purged from the design. This improves internal air quality as well as allowing biological materials to be returned safely to the biosphere.

A Positive Legacy

This book interprets the concepts of a circular economy into a simple set of design principles and business models that can be directly applied to the built environment. The case studies demonstrate how delightful spaces can be designed from reclaimed products and new materials that are good for the environment and for occupants, as well as saving on construction and materials costs. There are case studies that show how buildings that are designed to be adaptable to different uses (loose fit, long life) can retain value for longer by being more resilient to change in the market. There are examples of buildings that are designed to be completely demountable, allowing owners to flex their portfolio to changing market needs, alter their function and even move them to new sites in different configurations.

Chapter 13 and its case study (Park 20|20) demonstrates how a developer is reaping the rewards of moving to a more circular economy, by creating a virtuous circle that has reduced construction costs, increased rental yields, sidestepped depreciation and created buildings with residual value. The developer has formed collaborative partnerships with the whole supply chain that has promoted innovation over lowest-cost tendering and transparency of both cost and the make-up of each building component. Full accountability for the constituent parts of a building has created demonstrably healthier, more valued buildings. Most importantly, it leaves a positive architectural legacy for future generations.

01. What is a circular economy?

The essence of the circular economy lies in designing goods to facilitate disassembly and re-use, and structuring business models so manufacturers can reap rewards from collecting and refurbishing, remanufacturing, or redistributing products they make.

Ellen MacArthur Foundation

The circular economy is about keeping materials and resources in use and retaining their value, rather than consuming and disposing of them. To achieve this, products are designed to have longer lives, to be reused, remanufactured or reassembled instead of discarded. Or products have to be made from biological materials without any toxic chemicals so that they can be recycled or returned to the biosphere. Waste should be redefined as a resource to be exploited by local industries. As far as possible, new products should aim to design-out waste during their manufacture and at end-of-life, and ensure that residual waste can be reclaimed for reprocessing. A transition to a more circular economy means creating new industries and business models that focus on retaining the value of products and materials, and that redefine ownership by providing services rather than selling products. Ultimately, these new reuse and remanufacturing industries should be fuelled by renewable energy.

The current wasteful and inefficient 'linear economy' involves taking raw materials, manufacturing them into products and then throwing them away, often before they have failed or reached the end of their life. Some of the materials may be recycled, but valuable elements such as copper or aluminium are lost forever; it is currently impossible to separate copper or aluminium contaminants from re-melted scrap iron or steel.[1] This lowers the quality of the recycled iron or steel and means that the copper and aluminium are lost as a resource.

Some Definitions of a Circular Economy

A circular economy is an industrial system that is restorative or regenerative by intention and design. It replaces the 'end-of-life' concept with restoration, shifts towards the use of renewable energy, eliminates the use of toxic chemicals, which impair reuse, and aims for the elimination of waste through the superior design of materials, products, systems, and, within this, business models.

Ellen MacArthur Foundation, 2013[2]

The UK Government describes a circular economy as: *'moving away from our current linear economy (make-use-dispose) towards one where our products, and the materials they contain, are valued differently; creating a more robust economy in the process'*.

Environmental Audit Committee, 2014[3]

The European Commission proposes that we move to a more circular economy: *'This means re-using, repairing, refurbishing and recycling existing materials and products. What used to be regarded as "waste" can be turned into a resource. All resources need to be managed more efficiently throughout their life cycle.'*

European Commission, 2014[4]

This new economic model seeks to ultimately decouple global economic development from finite resource consumption. It enables key policy objectives such as generating economic growth, creating jobs, and reducing environmental impacts, including carbon emissions.

Ellen MacArthur Foundation, 2015[5]

The Origins of Circular Thinking

Walter R. Stahel is a Swiss architect and was an early advocate of the circular economy. He wrote a paper in 1982 called the 'The Product Life Factor' that proposed 'an economy based on a spiral-loop system that minimizes matter, energy-flow and environmental deterioration without restricting economic growth or social and technical progress'.[6] Figure 1.01 shows the 'self-replenishing system' from that original paper, which encapsulates many of the principles of a circular economy. In the diagram, the inner loop represents the reuse of goods and components (loop 1), the next loop represents repair (loop 2), the third is for remanufacture (loop 3), and the outer loop is for materials recycling (loop 4).

Figure 1.01: The self-replenishing system (product-life extension).

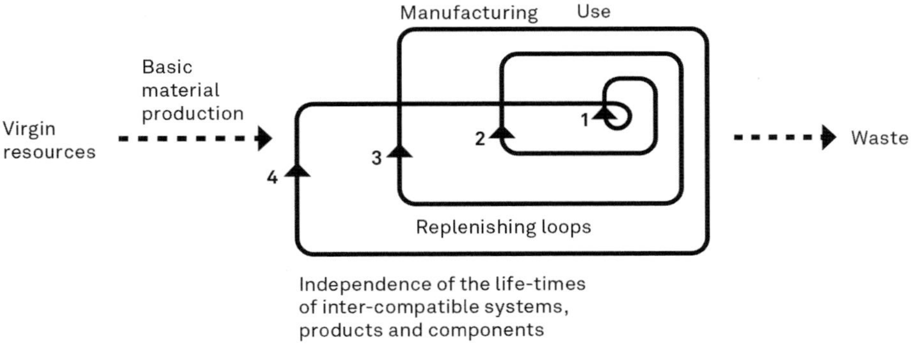

```
                    Manufacturing    Use
         Basic
         material
         production                1
Virgin   - - - - - ->         2
resources              3
                   4
                    Replenishing loops               - - - -> Waste

                    Independence of the life-times
                    of inter-compatible systems,
                    products and components
```

This shows that the concept of the circular economy is by no means new. The ideas are founded on several different schools of thought such as 'cradle to cradle', 'biomimicry' and 'industrial symbiosis'. It takes inspiration from biological cycles where nothing is wasted and waste is food.

Cradle to Cradle Thinking

The ideas behind the current interpretation of the circular economy are closely related to the Cradle to Cradle® design framework developed by William McDonough and Michael Braungart and explained in their seminal book *Cradle to Cradle*.[7] The Cradle to Cradle® philosophy goes beyond simply being about circles and creating closed loops; it is deemed by the authors to be more of a spiral of improvement.

Cradle to Cradle proposes the idea that instead of being 'less bad' to the environment, the aim should be to be '100% good'. It draws on the analogy of the cherry tree that 'makes copious blossoms and fruit without depleting the environment'.[8] It nourishes the soil, provides oxygen, absorbs carbon dioxide and provides habitats for many other organisms. The authors warn against 'eco-efficiency', as this can lead to buildings that have less access to daylight and views in a bid to reduce solar gains, or are tightly sealed without openable windows to reduce air leakage. Instead, they propose a building that has plenty of daylight and views, openable windows and night cooling strategies – an 'eco-effective' building.[9] They cite examples of buildings that they have designed that have delighted occupants, including a factory that has a brightly lit, tree-lined internal street with impressive staff retention rates.[10]

Cradle to Cradle includes a proposal that environmental 'regulation is a signal of design failure'.[11] There is no need to regulate to stop a pollutant entering the atmosphere if the need for that pollutant is designed-out from the start, or it can be fed back into the manufacturing process rather than being released. The authors worked with a furniture manufacturer to design a new fabric for upholstery that did not contain synthetic dyes and chemicals. Instead they designed a fabric that was free from toxic chemicals and used a biological material that could be composted at its end-of-life. This created a better product for users, the costs of disposing of the off-cuts from the factory were avoided and the effluent from the factory was as clean as the water going in. As an added bonus, the rooms required for hazardous chemical storage could be turned into recreation and work space, and protective equipment for the workers was no longer required.[12]

This idea applies directly to the construction industry: some fit-out contractors are exclusively using water-based paints for all their work, not because they are trying to reduce their environmental impact but because it avoids the need for storing hazardous chemicals and protects its staff from the fumes given off by solvent-based paints. This has knock-on benefits: the occupants will not have to breathe in toxic fumes and the waste will be less contaminated when the fit-out is torn out and replaced.

The authors of *Cradle to Cradle* have developed a certification scheme for products based on the principles they have developed. The scheme labels products based on a detailed assessment of the chemical ingredients in each material and their impacts on health and the environment. See Chapter 10 for more detail.

Biomimicry

Biomimicry is about mimicking the way that living organisms solve design challenges to produce sustainable solutions. The concept of the circular economy and cradle to cradle (C2C) philosophy both take their inspiration from living systems, where waste from one process becomes food for another organism and even poisonous substances can be processed and neutralised.

Inspirational examples of biomimicry in relation to the built environment are described in Michael Pawlyn's excellent book, *Biomimicry in Architecture*.[13] The book includes examples of how structural efficiency can be improved by mimicking the structure of bones and even ideas about how buildings can be grown. See Chapter 10 for more on biomimicry and how some of the philosophy can contribute to a circular economy.

Industrial Symbiosis

Industrial symbiosis is the study of how materials flow between industries and finding opportunities to use 'waste' from one industry as a raw material for another.

The National Industrial Symbiosis Programme (NISP) has been operating since 2003 and is a world first. Its aim is to inspire businesses to keep resources in productive use for longer. The NISP website (www.nispnetwork.com) includes many examples of industrial symbiosis in action, such as ceramic waste from the pottery industry becoming aggregate for a construction company, resulting in cost savings for both industries and reduced carbon dioxide emissions and waste.

Principles of a Circular Economy

A recent interpretation of a circular economy has been made by the Ellen MacArthur Foundation in a series of reports that include an evolved set of principles and compelling arguments for making the transition.[14] The reports include a diagram that captures and summarises the principles, as shown in Figure 1.02. The principles behind this diagram and how they apply to buildings are explained in this chapter and throughout this book.

From the very start, the aim should be to design-out waste and to think about products and materials as a precious resource that has to be preserved, rather than being wasted after it has been 'consumed'.

Figure 1.02: The circular economy.

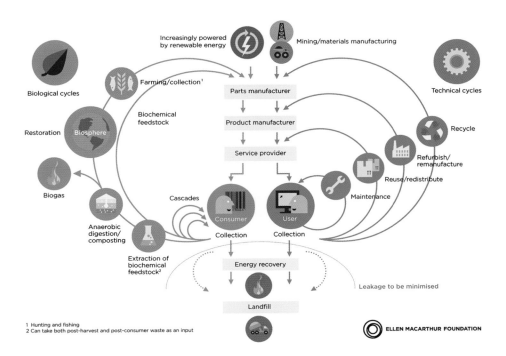

The diagram shows two distinct circles, to differentiate between 'biological materials' and 'technical materials'.

Biological materials, such as wood and natural fabrics, are used in products that are free of contaminant and toxins so that they can be returned to the environment at end-of-life. The diagram shows the potential for 'cascades', so a timber beam could be used in the building structure, then reused as a non-structural element, then formed into a fibreboard before being composted to generate biogas.

Technical materials, such as metals and plastics, are used in products that keep these materials in use for as long as possible and are retained within the 'technosphere' (technical cycle). This recognises that these materials are essential to the economy, that they are finite and that they should not be returned to the environment, as they may be toxic. Products made from technical materials should be designed to be durable and adaptable, allowing them to be upgraded or re-engineered, as necessary. These products should be designed to allow them to be disassembled at end-of-life. The components should be reused as far as possible and the constituent materials recycled as a last resort.

Buildings contain many complex components, particularly the plant and equipment used to service the building and the products used in the interior fit-out. As an example, fluorescent lamps contain many technical materials in very small doses, including rare earth phosphors, tungsten, nickel, zinc and aluminium. They also contain toxic substances such as mercury and chromium.[15] In the UK, lamps should be returned to dedicated recycling organisations once they are finished with, but in 2013 only 52.8% of the lamps put on the market were recycled.[16] This is a far higher collection and recycling rate than most consumer electronic goods, but it shows how difficult it is to close the loop. Once a lamp is purchased, it becomes the responsibility of the consumer to dispose of it correctly, and this does not always happen. This means that important and scarce metals and toxic substances are being lost or leached to the environment.

One of the ways to keep technical materials circulating is to shift away from the idea that consumers own products towards the concept that people are users who buy a service. When people purchase a product they are then responsible for its maintenance, repair and disposal. When users buy a service, the manufacturer has a vested interest in the whole life of the product and is ultimately responsible for its disposal. This should mean that manufacturers are incentivised to design products that can be repaired, upgraded and disassembled at end-of-life. This would allow them to keep the components and the materials in circulation for longer and significantly reduce the amount of material that is wasted. The idea of buying performance rather than products is discussed in Chapter 12.

Reusing Buildings and Components

Roughly speaking, half the embodied carbon in a building is tied up in the foundations and the structure.[17] Embodied carbon represents the carbon emissions released from extracting and manufacturing materials and elements, so it does not represent all of the environmental impacts from constructing a building. It ignores issues such as the scarcity of the material and its ability to be recycled at end-of-life. However, as a proxy for the resources needed to construct a building, it shows that reusing the foundations and structure of buildings makes a lot of sense.

There are models for designing buildings to be more flexible and adaptable to enable them to have a longer life (see Chapter 8). However, there are lots of reasons why buildings become obsolete and refurbishment is not always an option.

Once all the options for refurbishing the current building have been considered and the building is going to be demolished, then the next question should be: can the components be reused?

The scrap value of the materials in components is a tiny fraction of the value of the whole product. Reusing rather than reprocessing a component will also have far lower environmental impact. As noted by Bioregional, reusing steel sections can reduce the environmental impacts by 96% when compared to using new steel sections.[18] The difference between the scrap value and the value of the whole product is even higher for complex assemblies such as chillers or boilers.

There is additional cost and time associated with deconstructing buildings, rather than simply demolishing them, but this can be addressed through redesign and through rethinking the process of demolition and rebuilding. This is discussed further in Chapters 9 and 11.

Remanufacturing

Remanufacturing of plant and equipment is more common in other industries and it is an opportunity that has multiple benefits. According to a World Economic Forum report,[19] Renault has a remanufacturing plant in Choisy-le-Roi, France, that re-engineers different parts including water pumps and engines and then sells them at 50 to 70% of the original price, with a one-year warranty. The operation generates $270 million annually and employs 325 people.

In the construction industry, there is a limited market for refurbished boilers and there are, of course, architectural salvage yards and websites trading in a selection of valuable components, Belfast sinks, chimney pots and the like. In general, though, it is far more likely that building elements are scrapped and new ones are bought for the next project. An example of remanufacturing of furniture is described in Chapter 12.

Recycling

The outermost loop of the diagram is about enabling the recycling of materials back into the manufacturing process. This is the last resort once durable products have been through the other loops to extend their life as far as possible, retaining value and reducing the environmental impacts of reprocessing.

The design of the components should enable the materials to be readily separated for recycling and they should be clearly identified to allow them to be sorted and reprocessed. The materials should be kept as pure as possible, avoiding contamination from other materials, so they can be reprocessed into components of a similar grade.

The current practice is for materials to be downcycled rather than recycled, meaning that they end up as lower grade products. So concrete becomes secondary aggregates and solid timber becomes particle board.

Current practices and the potential for change are explored in Chapters 4 and 11, respectively.

Avoiding Leakage

The bottom of the diagram shows how incineration of waste (even for energy recovery) and landfill is considered 'leakage', as it is wasting valuable resources that leach out of the circle and are lost from the economy. In the UK, the construction industry has been successful in diverting waste from landfill with a high percentage being recycled in some way, but the residual waste is typically incinerated, meaning that it is lost forever.

Retaining Value

The ways that the circular economy model can create and retain value are discussed in the Ellen MacArthur reports and summarised in the Figure 1.03. To retain the value of products, the circular economy model requires that:[20]

- The inner circles are prioritised, as they use the least resources. For buildings, this means refitting and refurbishment over demolition and rebuild.
- There are more consecutive cycles, so buildings are refitted, adapted and refurbished instead of being pulled down. And, ideally, when they reach the end of their useful life, they can be disassembled into modules that can then be reassembled in a new configuration and location. This is explored further in Chapters 6 to 9.

01. WHAT IS A CIRCULAR ECONOMY?

Figure 1.03: Sources of value in a circular economy.

Prioritise the inner circles

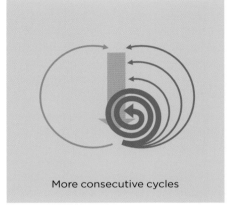
More consecutive cycles

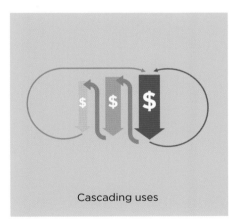
Cascading uses

Avoid contaminating materials

- There are cascading uses across industries with waste products from one industry becoming a useful product for construction and waste from buildings replacing the use of virgin materials in another industry. Biological materials should be cascaded through uses before being returned safely to the biosphere.
- The materials should be kept pure and uncontaminated to allow them to be reused or recycled into products of equal or better value or to be composted at end-of-life.

Applying all of these ideas to buildings will help to reduce the demand for raw materials while saving money, reducing the exposure to the volatile prices of resources and creating new, local industries.

Conclusion

The current idea and principles of a circular economy have evolved from many schools of thought into a coherent and highly compelling model that is starting to be applied by innovative organisations and individuals in many different industries. Chapter 5 shows how the principles can be applied to the built environment, while the case studies in this book demonstrate the practical implementation of the ideas in various projects.

A shift to a more circular economy means not only using fewer resources, but also creating and retaining value in buildings and their components. It should not inhibit design, but be used as a catalyst and inspiration for designing buildings that are a positive legacy for future generations. The shift will create new business models and industries that offer ways for organisations to create long-term relationships with customers and provide local employment opportunities from refurbishment and remanufacturing.

02. Why create a circular economy?

We have an uncertain future because our economy runs from a manufacturing perspective. We take a finite number of resources, we make something out of them and ultimately, at the end of their life, we throw them away.

Dame Ellen MacArthur

Population growth and the escalating demand for consumer products are driving up the demand for resources. It is predicted that, by 2030, three billion people who are currently living in poverty will join the middle-class level of consumption, which would create a corresponding surge in the demand for resources. According to the Ellen MacArthur Foundation, even the more conservative projections suggest that demand for natural resources will increase by at least a third.[1]

This increasing demand is creating price volatility and disruptions in the supply of essential raw materials, caused by conflicts, disasters or by countries restricting the international trade in raw materials, such as when China restricted the export of rare earth elements. This restriction created a sharp hike in prices of products that relied on these elements (e.g. in 2011, a price increase of over 300% was reported for rare earth phosphors used in fluorescent lamps).[2]

Environmental issues are the underlying reasons for many of the supply disruptions and climate change is causing more extreme weather events, adding to the problems. The Thailand floods reportedly led to a shortage of components for UK car manufacturers.[3]

Winning and processing resources is putting increasing pressure on the environment as these essential raw materials get harder to extract. Mining and drilling in more remote locations requires more energy, water and materials to access, and poorly controlled activities can result in the release of toxins that destroy farmland, fragile ecosystems and people's health.[4] Even pristine, precious ecosystems like the Amazon rainforest are mined for precious materials such as bauxite (for making aluminium). These mining activities lead directly to deforestation and the destruction of habitats and have other impacts, including an increase in soil erosion, pollution and the fragmentation of habitat caused by building access roads.

To compound the problem, it is becoming increasingly resource-intensive to extract fossil fuels, with offshore wells now having to be more than twice as deep as they were in the late 1990s.[5] This increases the demand for steel and means that more energy is required to extract a diminishing amount of fuel.

As noted by Allwood and Cullen in *Sustainable Materials: With Both Eyes Open*, most ores require the extraction of ten tonnes of rock to gain just one tonne of ore, and then the element has to be extracted from the ore. This extraction requires energy and uses chemicals, some of which are harmful.[6] Accidental or incidental emissions of these chemicals to soil, water and air can have serious and long-lasting impacts on the environment, surrounding species and local people. The extraction of raw materials has also been linked directly to conflicts in many parts of the world.

Copper is a good example of a key metal that is heavily used in buildings, including cladding, wiring, electronics, plumbing, heating and cooling pipework, heat exchangers in refrigeration, and so on. Most copper mining is located in arid areas, which puts increasing strain on scarce supplies. Water demand for copper mining is projected to grow by 45% by 2020 because of the decreasing copper ore concentrations.[7] It is estimated that there was about 7% copper in every tonne of ore 150 years ago, now it is more like 0.6%.[8]

An analysis of the water used to produce important materials identified those produced in areas of high water scarcity and that require high levels of water to produce.[9] These include common building materials such as iron, copper and aluminium, which are all mined in areas of water scarcity. It is expected that water constraints will limit production in Australia, South Africa, China, India and Chile. On top of this, climate change is expected to reduce water availability in all these countries. It is predicted that water availability will drive price increases for raw materials in the future.

The supply disruptions have prompted many countries to identify the materials that they consider 'critical' to their industries. A European Commission report on critical raw materials identifies 20 materials that are considered critical because of both their high relative economic importance and their high relative supply risk.[10] According to the EU report, the reason that the raw materials have a high supply risk is because a large share of the worldwide production comes from only a few countries.

The construction industry accounts for over half of the UK's resource use, by weight,[11] and is dependent on materials such as aggregate, cement, clay products, timber, lead, copper, glass, iron and steel for its structure and fabric. Buildings are now bristling with technology, with electronic components integrated into many products. This technology depends on the use of many critical materials, in small but essential amounts. Even the permanent magnets in motors for lifts, chillers, heat pumps and the like use rare earth elements such as neodymium iron boride to make them as energy efficient as possible.[12]

Benefits of a More Circular Economy

There are significant benefits in moving to a more circular economy. In the UK, the Waste Resources Action Programme (WRAP) estimates that 540 million tonnes of resources entered the economy in 2010, and nearly half of that is left to be managed as waste, with only half of that being recycled.[13] Even a small increase in the reuse and remanufacture would save tonnes of resources and save millions of pounds. The Chartered Institution of Waste Management estimates that reducing waste across Europe could save €72 billion per year and create over 400,000 new jobs in Europe.[14]

These are just the benefits of reducing waste in the system. One of the other key benefits of a more circular economy is to increase repair and remanufacturing of products; estimates suggest there is the potential to increase this industry by over £3 billion, with a corresponding increase in skills and employment opportunities.[15]

Businesses and industries that move to a more circular economy should reap important benefits if they embrace the concepts fully:

- Careful management of materials and resources will help to save money through closer links with the supply chain and a better understanding of resource risks.
- Companies can maintain control and even ownership of materials, so helping them to protect themselves against volatile prices and to provide more security of supply.
- There should be significant reductions in the cost of compliance with environmental legislation and the disposal of waste.
- Most radically, the move away from selling products towards providing a service will keep companies more in touch with their customers (and building occupants!) for longer, so providing more opportunities to sell-in more services and to better understand the needs of the customer.

These new business models can result in more repair, reuse and recycling of products. Rolls Royce has been applying a service-based model for many years by offering 'power by the hour' which covers 'full in-use monitoring, servicing, repair, remanufacture and replacement' of its engines.[16]

These new business models can have secondary benefits as well. When Philips started selling a lighting service, called Pay-per-lux, the leasing arrangement included incentives for Philips to closely manage the lighting controls to keep the energy consumption as low as possible (see Chapter 12 for a case study).

It's Already Happening

The Ellen MacArthur Foundation and McKinsey management consultants identify three trends that are contributing to a transition to a more circular economy:

> First, resource scarcity and tighter environmental standards are here to stay. Their effect will be to reward circular businesses that extract value from wasted resources over take-make-dispose businesses. Second, information technology is now so advanced that it can trace materials anywhere in the supply chain, identify products and material fractions, and track product status during use. Third, we are in the midst of a pervasive shift in consumer behaviour: a new generation of consumers seems prepared to prefer access over ownership.[17]

These trends are driving companies to change their business models and take advantage of the emerging markets and potential financial benefits. Some companies in the construction industry, notably carpet manufacturers including Desso, Interface and Shaw, are already implementing circular economy thinking and reaping the rewards, including designing carpets that can be readily recycled and creating leasing models that allow customers to buy a flooring service rather than owning a product. Desso, a major Dutch carpet manufacturer, has embraced cradle to cradle principles (see Chapter 1). According to Desso's CEO, the strategy has delivered increased profit margins despite the global economic crisis, indicating that customers are already willing to pay a premium for the greener product lines.[18]

Caterpillar has had a remanufacturing business since 1973. It manufactures and sells machinery and engines, including construction and mining equipment. It has handled more than 70,000 tonnes of remanufactured products in 2010 and been growing at a rate of 8-10%, according to the Ellen MacArthur Foundation[19] (see Chapter 12).

Conclusion

There are pressing reasons to reduce the demand for raw materials and to reduce the waste arising from the construction and demolition industries. The circular economy model creates financial benefits to those companies that adopt it as well as reducing the environmental and social cost of a consumer society.

The shift to a more circular economy is happening, with manufacturers seeing the benefits and offering new products and services, including some within the construction industry.

It is easier to see how a manufacturer will gain from applying circular economy principles, as it can benefit financially from maintaining more control over its supply of raw materials and reducing both waste and the cost of complying with environmental legislation. Consumer products are also relatively short-lived, so designing them for remanufacture or reprocessing will pay back in a few years, and new financial models can help to secure longer-lasting relationships with customers.

But buildings are not consumer products and they are not 'manufactured' in the same way as other products. This makes it more difficult to see how the circular economy model can be applied to buildings and the construction industry, with its complex and fragmented supply chain, long lifetimes and complete disconnect between construction and demolition. The next chapters explain the current situation and then show how circular economy principles really can be applied to the built environment.

03. Built to last?

We build to endure, to resist time, although we know that ultimately time will win. What previous generations erected for eternity, we demolish.

N. J. Habraken

When a new building is being designed and constructed, it is hard to imagine that it will one day be demolished. It is harder still to expend the time and effort to think about whether the building can be designed so that the materials and components can be reclaimed for reuse or recycling at the end-of-life.

Commercial buildings are often refurbished or demolished before their structure and fabric fails. The reasons for demolition are more likely to be related to changing land values, lack of suitability of the building for current needs or lack of maintenance of various non-structural components.[1]

Buildings such as laboratories become obsolete quickly due to rapid changes in technology or new research techniques, whereas high-quality housing can remain current and useful for longer.

Buildings that we value do not depreciate. Quite the reverse: they often have increasing value attached to them as they become older. The stock of listed buildings shows how the buildings that are really valued and loved are retained over their contemporaries.

On the other hand, Habraken points out that 'conservation may serve to freeze works of art in time, resisting time's effects. But the living environment can persist only through change and adaptation.'[2] Conservation can be criticised for stifling the natural

BUILDING REVOLUTIONS

complexity of the built environment and inhibiting its natural adaptation. This raises the question of whether the built environment should be continually refreshed and adapted, or retained and preserved as far as possible.

A paper by Forest Lee Flager contrasts Notre Dame Cathedral in Paris with the Ise shrine in Japan.[3] Gothic cathedrals across Europe are visited by millions of tourists a year to marvel at the feat of architecture and engineering, and the longevity of the buildings. Flager argues that the 'evolution of the Gothic cathedral is conveyed, at least in part, through the physical materials used in its construction. The years of effort required to bring the vision of Notre-Dame to completion are evident in the craftsmanship of every detail.'

This contrasts with the ancient Shrine complex in Ise, Japan. Known as the Ise Jingu, the shrines were originally built in the ninth century and have been rebuilt over 60 times since (see Figure 3.01) to reflect the principles of Shinto and Wabi-sabi, which hold that impermanence and ageing are intrinsic to all natural materials. The inner

Figure 3.01: The Mishine-no-Mikura shrine at Naiku, Ise (adjacent to the inner sanctuary), in Japan.

Figure 3.02: Taka-no-Miya shrine at Geku, Ise, in Japan showing the old and new shrines sitting side by side.

shrine, Mishine-no-Mikura, at Naiku, Ise, and the outer shrine, Taka-no-Miya, at Geku, Ise, are dismantled every 20 years and reconstructed on adjacent sites using the same design, with new materials, as part of a ceremonial ritual that lasts 17 years (see Figure 3.02). The buildings are reconstructed using traditional techniques before the originals are dismantled and their constituent parts distributed to other holy shrines and grounds for reuse.

The building is constructed from Japanese cypress, with a raised floor, a roof thatched with miscanthus grass and supporting timber pillars buried in the ground. In an article, Junko Edahiro states that the rebuilding ceremony is an important national event and explains:

> Its underlying concept – that repeated rebuilding renders sanctuaries eternal – is unique in the world. In the occidental way of thinking, creating something durable would normally involve building a structure with robust stones, bricks, and concrete. At this shrine, however, the structures are made exclusively from

wood and, by being rebuilt over and over again, can last forever. Also, in the process of rebuilding, the skills of shrine builders and craftsmen in various fields (carpentry, sacred treasures, apparel, etc.) are passed on from generation to generation.[4]

The shrines are built using elaborate joinery techniques with joints that are flexible but with enough strength to last without the use of nails, screws or adhesives. The connections make allowances for the shrine to be repaired and disassembled at the end of the 20-year cycle.[5]

Flager quotes the architect Kisho Kurokawa to explain the different philosophies:

> We have in Japan an aesthetic of death, whereas you [Westerners] have an aesthetic of eternity. The Ise shrines are rebuilt every 20 years in the same form, or spirit; whereas you try to preserve the actual Greek temple, the original material, as if it could last for eternity.

If buildings are designed to be permanent, then they have to be of a very high quality and be greatly valued by future generations to survive. The alternative is to think of a building as a fleeting, temporary structure that will ultimately be destroyed to make way for the new. This means buildings can become disposable consumer goods, which is faintly absurd when they demand so much resource to build and generate so much low-value waste during refurbishment, refit and demolition.

Building Obsolescence

The value of a building is a complex interaction between factors such as:

- market forces
- regulations
- technology changes
- build quality, and
- physical deterioration.

The reasons for a building becoming obsolete and then being demolished may have nothing to do with its physical condition; it may just be because its original function has become redundant. Warehouses in many of the docklands in the UK became obsolete when goods were brought in by containers rather than being offloaded by hand from small ships. Many of these warehouses were ideal for adaptation into other uses, ranging from offices and housing to markets and museums. For example, the original Georgian warehouses in West India Docks, London, now house the Museum of London Docklands, restaurants, shops and apartments.

Research shows that quality and characteristics of the building are a better explanation of depreciation than the age of the building.[6] Research by Baum and McElhinney notes that building lives are getting shorter, but that 'depreciation is not forever: depreciation for older property is lower than depreciation on new property. Better underlying value is available in older buildings.'[7] The problem comes when the value of the building depreciates to the point that it is close to the value of the land; at that point, it is likely to be demolished and the land redeveloped.

So build quality can be an important way to retain value as poor build quality will build up the cost of adaptation,[8] making it more likely that buildings will be demolished.

Conclusion

Designing buildings to be more flexible and adaptable should allow them to retain their value for longer by avoiding functional obsolescence. Alternatively, buildings can be consciously designed for a short life and with the ability to be disassembled so that modules can be redeployed in new buildings or so that the materials can be separated and reclaimed for further use.

Either way, designing buildings that are easier to refit, refurbish or dismantle helps to reduce the demand for raw materials and to reduce the amount of waste arising. The best way to design for disassembly and reuse is to understand what happens at the end-of-life, as explained in the next chapter.

04. Starting at the end

The construction sector has a high raw material dependence and handles materials with high intrinsic value, while generating significant volumes of waste.

Ellen MacArthur Foundation

Buildings are a complex assembly of materials that are designed for construction and use, with little or no consideration of how they will be adapted and refurbished or demolished at end-of-life.

Consequently, any changes to the building during its life or at its demolition create tonnes of waste that are typically downcycled into lower grade products rather than being reused or recycled. That is aside from the fact that buildings are often demolished before the building elements have failed.

Great strides have been made in increasing the amount of construction waste that is diverted from landfill – often over 90% in some countries. But this raises the question: where does the waste go now that it is not being buried in the ground?

Current Demolition Practices

Demolition practices have, in general, moved away from the salvaging and reclamation of materials towards more mechanised demolition and reprocessing of materials. The demolition process has become increasingly mechanised due to the requirement to rapidly clear the site, to improve health and safety standards and to reduce the cost of demolition by using less manpower and more machines. Deconstruction is still undertaken on constrained sites where heavy machinery could affect adjacent structures or cause nuisance to neighbours, and is also used in refurbishment projects if more time and budget is available. Deconstruction can take twice as long and manual methods require more labour and require more access on site, which raises further safety risks.

Reclaimed materials have to be sorted, cleaned and indexed, all of which takes time and effort. They then have to be carefully stored and a buyer has to be found. On the other hand, the scrap value of materials is high and the returns are immediate, which helps with cash flow during the demolition contract.[1] It is clear from this why many demolition contractors choose recycling over reclamation.

The 'soft strip' of a building is usually the first activity on site after the initial planning stages and the services have been disconnected. The soft strip includes the identification and removal of any hazardous wastes (e.g. asbestos and fluorescent tubes), followed by a first pass to remove fixtures and fittings and then a second pass

Figure 4.01: Scrapped raised floor tiles. Could these have been reclaimed?

Figure 4.02: Demolition site showing how a building is torn down, leaving a pile of mixed waste.

to remove redundant services. Non-structural elements, such as partition walls, may also be removed at this stage.[2] Metals in the building services and internals, including aluminium, steel and copper, are removed as part of the soft strip and sold for reprocessing. Even complex and valuable building services equipment, such as chillers, pumps and boilers, are often sold as scrap rather than considered for reconditioning and reuse.

Once the soft strip is completed, the building fabric is then demolished as quickly and efficiently as possible. Steel beams are cut into manageable lengths using cutting torches or shears. Concrete is broken up and the reinforcing bars are segregated for recycling.

The materials arising from the demolition are either segregated and reprocessed on site or are sent to Materials Recovery Facilities (MRFs), depending on the constraints of the site and the processes involved. Concrete, blockwork and hardcore materials can be segregated and crushed on site for use as aggregate. High-quality bricks laid in sand lime mortar may be reclaimed, but the majority are crushed. Timber and timber products, plastics, plasterboard, insulation, ceramics and packaging are typically sent to MRFs for segregation and ultimately for reprocessing.

What Happens in an MRF?

In the UK, MRFs process the majority of mixed waste from construction sites. These processing facilities are cavernous industrial buildings containing huge piles of mixed waste and queues of trucks coming to replenish the heaps as they are sorted (see Figure 4.03).

Figure 4.03: Construction and demolition waste at an MRF prior to sorting.

Excavators are used to sift out the bulky waste, including mattresses, large metals and composite products such as electrical equipment. Inert waste, such as aggregate, may be diverted to different facilities for processing. Some waste may be shredded to make it easier for magnets and other automatic machinery to extract materials. The waste is then dropped onto an elevated conveyor belt. The metals are picked out with magnets and eddy current separators are used for the non-ferrous metals. Soil and other residue are filtered out using a vibratory screen. The level of mechanisation varies, but in a typical MRF the waste passes to the picking area where a gang of staff (known as 'pickers') grab and throw materials into the relevant bays below the conveyor. The pickers pull out wood, particle board, plastics, any remaining metals and cables as the waste travels along the conveyor. Some facilities also sort hard plastics into broad polymer groups. After the picking line, another vibratory screen or trommel shakes out any remaining inert waste such as aggregate and concrete. At the end of the conveyor, a new heap of residual waste is formed, representing around 20–40% of the material that was delivered to the site.

Any untreated solid timber is downcycled into panel products or wood fuel pellets. The composite and treated timber is chipped and used as Refuse Derived Fuel. Soft plastics are recycled back into other plastic products. The metals are shredded and smelted into, often, inferior grade metals. The hardcore is crushed and screened to be reused as secondary aggregate.

There are some waste materials that cause problems in MRFs. Composite products are difficult to process and will often end up as residual waste. Plasterboard that is correctly separated at construction sites can be sent for reprocessing, and can be turned into new plasterboard if it is salvaged as part of a take-back scheme. Otherwise, it can end up as treatment for farmland or compost. Any plasterboard that is broken up and mixed with other waste has to be treated separately and sent to dedicated mono cells at landfill sites. This is because plasterboard gives off hydrogen sulphide if it is disposed of in conventional landfills.

The majority of the residual waste becomes Refuse Derived Fuel by being shredded, baled in plastic and shipped out to incineration plants in continental Europe, where it provides heat for buildings.

So, a 90% diversion from landfill really means that around 80% of the diverted materials (at best) are recycled, or more accurately downcycled to a lower grade material, and the rest is incinerated.

Conclusion

The demolition industry is moving towards increased mechanisation and away from deconstruction to save time and money and to ensure the safety of workers. Selling high value materials for reprocessing provides a faster return than reclaiming them for reuse. This means that building materials from demolition activities are unlikely to be salvaged for reuse without a radical rethink. The demolition and waste processing industries are simply playing the hand that is dealt them. To change, they need better cards. And that means going back to the beginning and rethinking the design of buildings.

The current building design and construction practices are creating a legacy for future generations that will make it increasingly difficult for them to reclaim valuable, uncontaminated materials in an increasingly resource-constrained world.

To bequeath the next generation with buildings that can become 'resource banks' in the future means that designers and constructors have to reverse the current trends in the industry and embrace the principles set out in this book.

05. Circular economy principles for buildings

Look at a building, there are hundreds of tonnes of valuable materials in those buildings. The residual value of those buildings and materials is a negative one – we have demolition cost and the valuation of commercial real estate is based on discounted cash flows. No-one takes [into account] the materials because [buildings] are not designed to be material banks.

Coert Zachariasse

Applying circular economy principles to the construction sector and buildings offers huge potential rewards.

Buildings can be designed to have a positive, enduring legacy by making them more adaptable and by ensuring that valuable materials and components can be reclaimed and reused at end-of-life. Ensuring that buildings can be disassembled provides the opportunity for them to be redeployed in new places or for new uses, and allows components to be salvaged and reused or remanufactured. This, in turn, reduces dependence on raw materials for construction while salvaging and remanufacturing creates local employment. Declaring and understanding the ingredients that make up materials and components will help to ensure that biological materials can be safely returned to the biosphere and technical materials can be reclaimed for reuse within industry. There is also the added benefit that the use of pure materials with the contaminants designed-out helps provide better environments in which people can live and work. Figure 5.01 summarises the principles of a circular economy when applied to buildings.

Figure 5.01: Applying circular economy principles to building design.

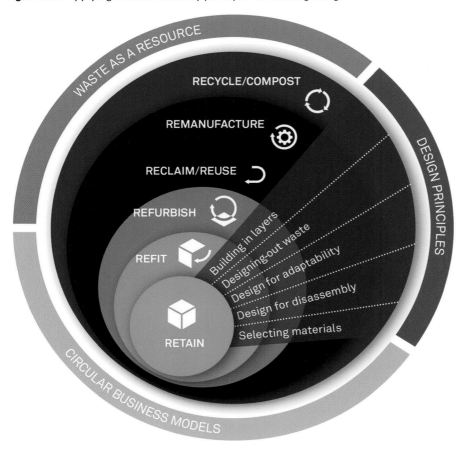

Concentric Circles

The nested circles show the hierarchy with the three inner circles being the most desirable. Retaining the existing building is the most resource-efficient option, followed by refits and refurbishment of existing buildings, as this retains the most resource-intensive parts of the building. For the three outer circles, the priority is to reclaim or remanufacture components, with the last option being to disassemble them to recycle back into new products or return the materials to the biosphere. This hierarchy underpins the design principles covered in this book.

05. CIRCULAR ECONOMY PRINCIPLES FOR BUILDINGS

Design Principles

The five segments overlaid on the circles show the design principles that can be applied to reduce waste, extend the life of the building and enable the reclamation of materials at end-of-life. These design principles are covered in Chapters 6 to 10 and summarised below:

- The idea of 'building in layers' recognises that the various elements of the building have different lifespans and should, therefore, be independent to allow different layers to be peeled off and replaced or salvaged without damaging the adjacent layers (see Chapter 6). This helps to create buildings that are simpler to maintain, flex or adapt, and it allows the components to be more readily reclaimed at end-of-life.
- 'Designing-out waste' means prioritising the refit and refurbishment of existing buildings, as this preserves the most resource-intensive elements of the building. Using reclaimed materials and remanufactured products along with leaner designs will reduce the demand for raw materials. Lastly, using modern construction techniques can avoid waste arising on site (see Chapter 7).
- 'Design for adaptability' means that buildings can be retained for longer. Designs have to consider how the building could be converted to other uses and how that affects the structural design and internal reconfiguration. Buildings that have proved to be adaptable have some common features, as discussed in Chapter 8, but it often seems to come down to designing buildings that are valued by people, as well as the plan form, layout and structure.
- 'Design for disassembly' and reuse allows components, or even whole buildings, to be reused. Chapter 9 includes an example of a building that has had three lives, being disassembled and reassembled on different sites for different purposes. This means that buildings can become assets that are independent from the value of the site and retain their value for longer. Equally, materials will have more value if they can be extracted, turning buildings into 'materials banks'.
- Lastly, when selecting building materials and products, the constituent elements have to be known and they have to be split into biological and technical materials to allow them to be either returned to the biosphere or kept in an industrial loop of recycling or reuse (as explained in Chapter 10). The lifespan of the component should be matched to the materials selected to avoid wasting valuable resources when products are replaced long before their technical lifespan. Technical materials that are difficult to reclaim or recycle at end-of-life can be replaced with biological materials that can simply be returned to the biosphere.

New Business Models

The outer ring in the diagram represents the underlying models that can be applied to enable a more circular economy across the buildings sector.

Turning the idea of waste on its head and treating it as a resource is a fundamental principle of a circular economy. For the construction industry, this means creating a market for salvaged products and materials by using them in the design and refurbishment of buildings in preference to new ones. This, in turn, means that a detailed inventory of the constituent parts of buildings has to be created so new markets for these materials can be found before the building is stripped out or demolished. Chapter 11 explores the idea of turning buildings into 'materials banks' where materials are deposited and then withdrawn at a later date. There is also the potential to sell salvageable goods well in advance of demolition to create an incentive for owners and contractors to have more time to reclaim elements of the building.

The new business models move away from the linear economy approach of purchasing products, consuming them and being responsible for their maintenance, upgrade and disposal. Instead, customers and manufacturers develop longer-term relationships, with customers being able to purchase performance instead of products. This means that manufacturers have a vested interest in designing products that can be maintained, upgraded or recycled and it helps them to secure a supply of components and materials in the future. Chapter 12 looks at these new models.

Conclusion

The overarching philosophy is to put thought into the future destiny of the building and the legacy that it leaves for the next generation.

Buildings are often considered as an endowment for future generations, but all too often they become a liability that has to be demolished at a cost. Commercial buildings often depreciate over time until they have no residual value. Buildings that are valued by people do not depreciate and can often appreciate over time. By thinking about the potential future life of a building and learning lessons from the buildings that have endured, designers can create buildings that are more flexible and adaptable, giving them a longer life. Alternatively, designers can deliberately design for a short lifetime and ensure that the elements of the building can be readily disassembled and reused at end-of-life. There is even the potential to design buildings that can be demounted and reassembled in new locations and reconfigured for new uses. This is particularly relevant to organisations that experience rapid change in customer demand.

By thinking more about the whole life of the building, designers can create a lasting positive legacy for future generations, gifting them either adaptable buildings or giving them readily accessible materials and components with which to create new buildings.

06. Building in layers

Our basic argument is that there isn't any such thing as a building. A building properly conceived is several layers of longevity of built components.

Frank Duffy

The lifetime of different elements of a building can vary from well over a hundred years down to a matter of months, or even weeks. The structure and fabric of a building, when made from durable materials such as stone, slate, brick and concrete, has been proven to last for hundreds, if not thousands, of years. Newly installed components can fail, prematurely, in a few weeks or be torn out when they do not meet the new occupant's needs or expectations. This calls for a different approach to designing structure and fabric as compared to those elements with a shorter life.

There has to be a clear delineation between the elements with different lifetimes, so sealing electrical wiring into walls or encasing them behind durable finishes means that there will be disruption and waste when they need replacing or moving. Equally, some products may have built-in obsolescence with inaccessible components that fail prematurely.

Frank Duffy proposed a layered approach to building and identified four layers:

1. Shell
2. Services
3. Scenery, and
4. Set.

Figure 6.01: Shearing layers of change.

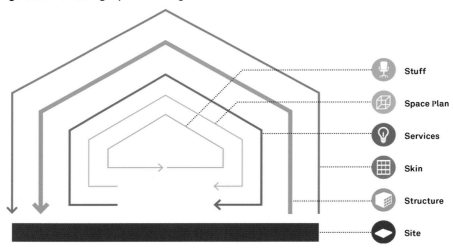

The Shell is the structure; the Services are pipes, ducts and wires; the Scenery is the internal fit-out; and the Set is furniture, fittings and equipment. Each of these has a different lifespan with the Shell lasting the longest (the life of the building) and the Set lasting only a few months, depending on the occupants.

Stewart Brand expanded and adapted this model in his famous book, *How Buildings Learn*,[1] into six:

1. Site
2. Structure
3. Skin
4. Services
5. Space Plan, and
6. Stuff

(see Figure 6.01).

In Brand's categories, the structure and the skin are separated, which accords with the Open Buildings philosophy (see Chapter 8). The Site is the geographical setting; the Structure is the load-bearing elements, such as the foundation and skeleton; the Skin is the exterior surface, such as the façade; the Services are the circulatory and nervous systems of a building, such as its heating plant, wiring and plumbing; the Space Plan includes walls, flooring and ceilings; and Stuff includes lamps, furniture, appliances and ICT.

This layering approach is invaluable when considering how to approach the different elements of a building.

Considering the lifespans of different layers and components in the building has benefits at each stage in a building life. During construction, it can help with the sequencing of tasks, and broadly follows the construction process (e.g. structure is followed by skin, then the internal fit-out). During the operation of the building, careful separation of components dependent on actual lifespan should help to make the short-lived components more accessible for replacement and provision should be made to access components that require regular maintenance, repair and upgrade.

During refits or refurbishment, the ability to peel off layers and apply new ones ensures that the neighbouring layers are undamaged. In one way, the layering approach needs some refinement: it is very likely that building services will need to be maintained or replaced before the finishes (Space Plan) are changed, and finishes such as tiling could have a far longer lifetime than the services buried behind them.

Creating buildings that are more flexible and adaptable to other uses also draws on the idea of layering the building by ensuring that the primary structure is independent from the secondary structure, allowing for more interventions.

Finally, the use of independent layers helps to enable disassembly at the end-of-life of the building by allowing each element to be removed independently.

Layers of Complexity

Figure 6.02 shows how the different layers can be separated based on the intended life of each element.

The Shell

The structure of the building has the longest potential lifespan and is the limiting factor in adapting a building to a new use. If the façades and the internal configurations are independent of the structure, this should enable adaptation. Chapter 8 discusses the principles for designing for adaptability. This includes designing-in some level of redundancy in the structure and providing generous floor-to-ceiling heights, providing carefully positioned service cores and the capability for new service runs to be installed between floors or across the floor plate.

Ideally, the structure and the fabric need to be independent to allow parts of the façade to be replaced or a whole new façade installed. The ability to exchange, say, a glazed unit for an inlet louvre or a fully glazed façade with a solid element would make a building more adaptable to different uses.

Within the fabric layer, there are further sub-layers to consider, especially when designing for disassembly (see Chapter 9). This includes ensuring that the different layers of the fabric of the building can be easily separated at end-of-life. In particular,

BUILDING REVOLUTIONS

Figure 6.02: Layers of a building and circular economy principles.

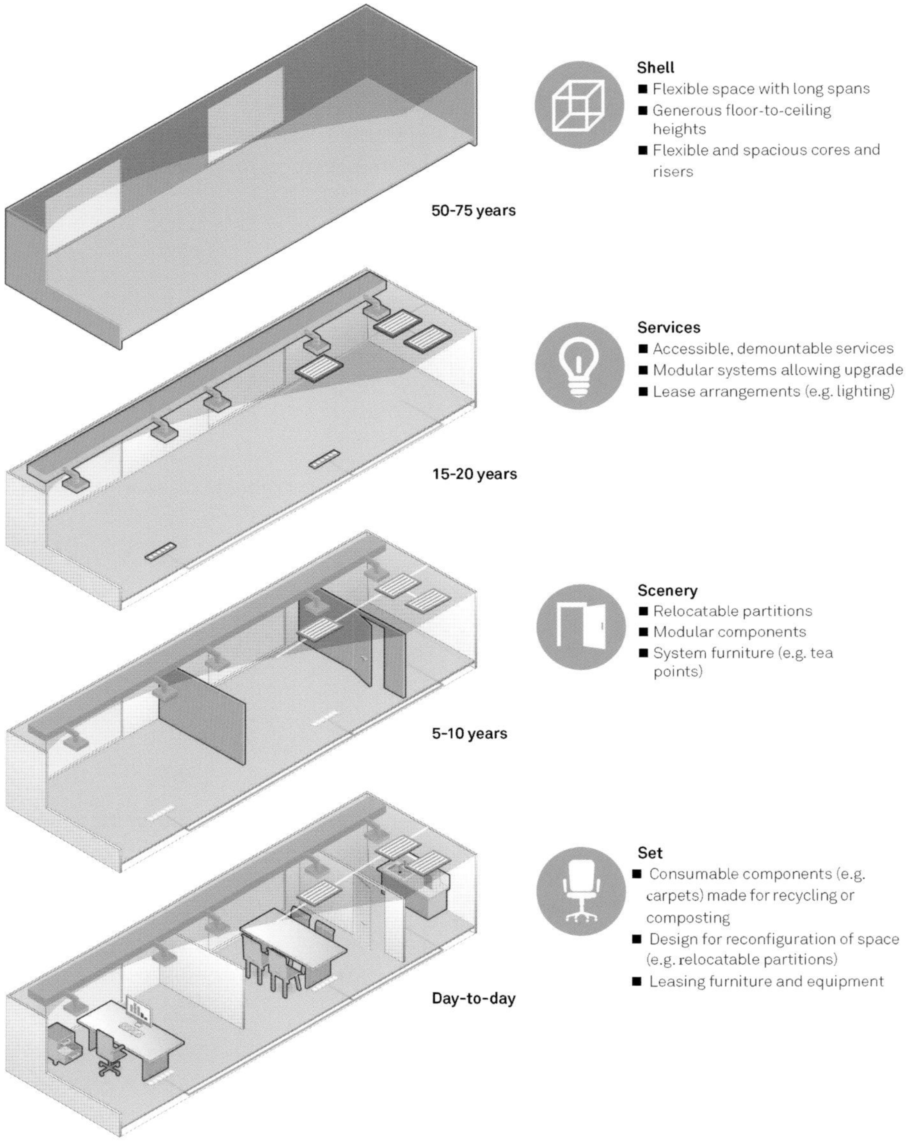

Shell
- Flexible space with long spans
- Generous floor-to-ceiling heights
- Flexible and spacious cores and risers

50-75 years

Services
- Accessible, demountable services
- Modular systems allowing upgrade
- Lease arrangements (e.g. lighting)

15-20 years

Scenery
- Relocatable partitions
- Modular components
- System furniture (e.g. tea points)

5-10 years

Set
- Consumable components (e.g. carpets) made for recycling or composting
- Design for reconfiguration of space (e.g. relocatable partitions)
- Leasing furniture and equipment

Day-to-day

elements that use composite constructions with integral insulation have proved hard to separate for reuse, or even recycling, at end-of-life.

Services

Commercial buildings often make provision for accessing services through access panels, suspended ceilings and raised flooring system. However, services are still often buried in walls, tangled up with the structure or trapped in the building with no provision for removal and replacement. In traditional brick and stone buildings the structure and the façade are combined and the services are often woven through all of the elements.

Using the layered approach also helps to make the building easier to maintain, as the services will be more accessible for repair and maintenance. The design strategy should identify the elements that need maintenance or upgrade and ensure that they are accessible and have components that are durable and repairable.

Scenery

Scenery, such as partition walls, floor finishes, built-in cabinets, etc., will have a shorter life than the other elements of the building. Refits or reconfigurations of internal spaces generate a considerable volume of waste that is typically downcycled. Scenery should be designed to be reconfigured, reused, relocated or recycled (rather than downcycled) or it can be made from biodegradable materials that allow it to be composted at end-of-life.

Set

The consumable components, such as carpets and furniture, can be designed with a shorter life and made of biological materials that can easily be recycled or broken down at end-of-life. Several carpet manufacturers already have ranges of carpets that meet these criteria, and furniture manufacturers are designing closed loop products. For example, chair manufacturers are designing their products so that they can be readily recycled and it is now even possible to obtain cardboard furniture. Some manufacturers are offering leasing arrangements that mean customers can purchase a service (e.g. lighting or seating) rather than a product. This gives customers more flexibility and allows the manufacturer to retain ownership of the products (see Chapter 12 for more information and some examples).

Conclusion

The principles of the circular economy cannot be applied, wholesale, to buildings as they are a complex collection of products, all with different lifespans, purposes and demands.

The idea of designing in layers helps to unlock this tricky problem by allowing designers to approach each element of the building using different rules. So, the structure and fabric can be made to be adaptable and over-engineered to last a lifetime, while the internals will be fickle and have to be designed to be reusable or compostable. Similarly, the idea of leasing a structural beam may be far-fetched, but leasing furniture instead of purchasing it, makes complete sense.

Keeping each of the layers independent allows the structure to be retained when upgrading the fabric and the building will be easier to disassemble at end-of-life so that the components can be reused, remanufactured or recycled.

07. Designing-out waste

Waste does not exist when the biological and technical components (or 'nutrients') of a product are designed by intention to fit within a biological or technical materials cycle, designed for disassembly and refurbishment.

Ellen MacArthur Foundation, 2013

A circular economy is not just about designing-out waste, it is fundamentally designing-out the concept of waste. This means that designers have to think about the whole life of the building from the decision to build new or refurbish through to the eventual demolition or deconstruction of an obsolete building.

Therefore, the idea of designing-out waste means avoiding creating waste in the first place, and looking for opportunities to turn waste from other places into a resource. For buildings, this includes:

- refitting and refurbishing existing buildings rather than building new
- designing-out waste arising during construction
- using reclaimed materials and components in design
- applying lean design principles to reduce demand for resources and associated waste.

This chapter covers each of these ideas, along with case studies that show how different elements of these principles can be applied.

Refit and Refurbishment

The greenest building is the one already built.

Attributed to architect Carl Elefante

When deciding whether to refurbish an existing building or to rebuild, the constraints of the existing building are often cited as the reason for demolition. Typical issues are the floor-to-ceiling heights, riser space or the floor layout. There is also often a financial incentive to build new, as this can increase yields and, in the UK, the tax regime offers little incentive to developers to refurbish existing buildings.[1] On the other hand, there are many situations where buildings are refurbished instead of being demolished and the designers work around these constraints. In the UK, some developers have built their business models around refurbishing existing buildings into attractive places to live and work.

The Tea Building and the 'White Collar Factory', London

For Derwent London, breathing new life into old buildings is a central tenet of its business model. It aims to reuse as much of the fabric of the original building as possible to reduce resource use, time on site and to save money. The Tea Building in

Figure 7.01: The Tea Building in Shoreditch, London.

07. DESIGNING-OUT WASTE

Shoreditch, London, is a great example. Built in the 1930s as a bacon factory for Allied Foods' Lipton Tea Brand, it was used as a tea-packing warehouse for most of its life. Derwent London converted it into a series of individual office units by keeping the structure and façade, refurbishing the windows and installing new main plant and services (see Figures 7.01 and 7.02).

Figure 7.02: The Tea Building in Shoreditch, London.

Much of Derwent London's portfolio has been carefully selected to have generous floor-to-ceiling heights, good day-lighting and exposed thermal mass to help regulate temperature.

The Tea Building, along with the Johnson Building and Horseferry House, is the inspiration for its latest design concept, dubbed the 'White Collar Factory'. According to this concept, Derwent London would avoid wasting resources on fitting-out the building in case incoming tenants wanted something different. Instead, it would provide a minimal, efficient initial servicing strategy giving tenants a high level of flexibility about how they want their spaces to be fitted-out and serviced. So meeting rooms and cellular offices could be added as pods to be plugged into the concrete core cooling system embedded in the ceiling slabs and acoustic partitions, along with any other elements, as required. To assist prospective tenants, Derwent London has arranged and costed a range of options to help the decision-making process. Figure 7.03 shows one of the fit-out options.

These examples show that considering the refurbishment of an existing building can produce a building that is as attractive as a new building and can even be more appealing to occupiers.

Figure 7.03: 'White Collar Factory' fit-out visualisation with mezzanine.

55 Baker Street, London

Sometimes it takes a visionary design team to propose a solution when the only option appears to be rebuilding. In the redevelopment of 55 Baker Street, it was assumed that the huge 1950s office building would have to be demolished to allow the site to be transformed into a new urban centre. Instead, Expedition Engineering (the structural engineers) and Make Architects proposed a solution that would preserve and enhance the existing building as far as possible. The developer, London and Regional Properties Limited, immediately grasped the benefits of the refurbishment option over the new build option and appointed the team to implement the design.

The floor layout of the existing building was constrained by the eight full-height vertical circulation cores that inhibited the creation of the open plan floor plates typically demanded by commercial tenants. The structural engineers proposed retaining the majority of the existing reinforced concrete frame and the targeted demolition of the cores to allow the creation of the open floor plates. The cores were replaced with new concrete infill structures to provide lateral stability.

New building structures were inserted into the original 'H block' layout to create a design that includes three atria and infill blocks that created additional floor space and connected adjacent wings of the original building. The new glazed façades provide a new, coherent face to the building along the front elevation (see Figure 7.04).

Figure 7.04: 55 Baker Street: before and after refurbishment.

Figure 7.05: Steel transfer structure in entrance.

A steel transfer structure was installed to allow the removal of 12 existing reinforced concrete columns supporting seven floors above, opening up the entrance and reception area (see Figure 7.05).

This project shows how new life can be breathed into a building that is destined for demolition. Around 70% of the original building structure was reused and the upgrade took a year less than a rebuild solution.[2]

Designing-out Waste on Site

The amount of waste generated on construction sites can be reduced considerably by changing the way that buildings are designed and constructed. The most effective way to reduce waste arising on site is to move more of the construction activities off site and make the work on site more about assembling components than the cutting and shaping of materials.

Even when using traditional construction techniques, there are opportunities to design-out waste. Examples include using modular components (e.g. doorsets rather than doors) and coordination of the structural grid with the external cladding and internal finishes and partitions. The use of 3D design software can be invaluable in ensuring that designs are coordinated and to help to avoid wastage on site when components do not fit. Designing to match the standard sizes of sheets and panels can help to avoid off-cuts and the use of smaller board sizes can help to deal with complex geometries.

The carpet tile manufacturer Interface® has used product design to reduce site waste. It has created a 'Random Design' carpet tile that allows the tiles to be laid in any direction. This reduces cutting waste as well as making installation quicker and easier.

It is estimated that more than one million tonnes of plasterboard waste is produced in the UK each year and wastage of 10–35% often occurs on site through wasteful design, off-cuts, damaged boards and over-ordering.[3] Waste also arises from the removal of plasterboard during refits and soft strip prior to demolition. In a WRAP case study of the Tate Modern, the use of fair-faced or rendered finishes reduced the dry lining costs by £43,240 and avoided four tonnes of potential waste.[4]

These techniques will certainly reduce waste arising on site, but when considering a more circular economy there are additional fundamental questions about whether the materials are needed at all. In particular, can reclaimed materials be used instead and can the need for the materials be designed-out completely?

Reusing Components and Materials

Moving to a more circular economy does not mean that each industry has to create closed loops for its components and materials. Materials designated as 'waste' from one industry can become a valuable resource for another; this is the principle of industrial symbiosis (as explained in Chapter 1).

Chapter 4 highlighted that the amount of materials salvaged from demolition sites is dropping and there has been a shift away from using reclaimed materials in the UK. However, there are architects and interior designers who are embracing the opportunities to incorporate reclaimed materials into their designs.

These opportunities have to be considered right at the start of the project. The ICE Demolition Protocol proposes carrying out a pre-demolition audit of existing buildings and infrastructure on the site to identify materials that could be reused in new buildings.[5]

When the Demolition Protocol was implemented on the London 2012 Olympic Site, a wealth of materials was identified for potential reuse and some reused on site. Over 220 buildings had to be demolished on the site of the park along with walls, bridges and roads. The Olympic Delivery Authority set an ambitious target to reuse or recycle 90% of the material (by weight) arising through the demolition works, prioritising reuse on site over recycling. Pre-demolition audits were carried out on all the buildings and infrastructure and the results of these audits were compared with the demand for materials for the new development. The cost of reclaiming all the materials was deemed too high, so the Olympic Delivery Authority focused on those components that could be reused locally. This was mainly the yellow stock and Staffordshire Blue bricks, the granite kerbs and sets/cobbles, roof tiles, street furniture and concrete kerbs and paving.[6]

The natural instinct is to try and source materials from within the construction industry, but industrial symbiosis promotes the idea of creating links between different sectors with waste from one become a resource for another. Some innovative designers are looking outside the construction industry for sources of materials for fitting-out buildings.

Huckletree, London

Grigoriou Interiors is a London-based interior design company that specialises in sustainable interiors. When commissioned to design the new 'shared workspace' offices for Huckletree, the design director Elina Grigoriou found a unique source for reclaimed materials: salvaged materials from film and television sets like the *Fast and Furious* film and the BBC's drama series *The Hour*. Through an informal network of friends and relatives, Elina discovered Dresd, a company that salvages scenery, props, furniture and materials from TV and film production companies and resells or hires them back to the industry for a fraction of the cost. Grigoriou Interiors was able to use Dresd's database of materials and products to inform a large part of the design of the new office space and to radically cut down on the use of raw materials in the fit-out.

The benches were designed to be made from old scaffolding planks and poles (see Figure 7.06). The planks were sanded down enough to be smooth, but to still retain the aged patina of the wood. Wooden scaffolding ladders were used to form screened off areas and shelving, along with more scaffold planks. The timber finishes have no Volatile Organic Compounds (VOCs) to help create a healthier internal environment, as well as ensuring that the timber can be reused or returned safely to the biosphere at end-of-life.

The partition walls for the Skype booths on the first and second floors, and the screens at the end of desk rows, were made from reclaimed glass, with the exact dimensions being dictated by what was available (see Figure 7.07). The soft seating area lampshades were made from laser-cut cardboard, and the 'Huckletree' sign over the entrance used reclaimed plywood laser-cut to form the letters and painted. Internal signage was also made from reclaimed timber, while graphics or user information was creatively integrated as artwork directly on walls or painted reclaimed ply surfaces.

07. DESIGNING-OUT WASTE

Figure 7.06: Huckletree interior showing reclaimed wood benches.

Figure 7.07: Huckletree interior showing reclaimed glass screens.

The desks, furniture storage and task chairs were all sourced from Ahrend, a cradle to cradle certified company. This certification works to ensure that the materials in the products are known and that they are not toxic to human health. They are also designed for disassembly, allowing them to be returned to the manufacturer for remanufacture, with the biological materials safely returned to the biosphere.

The whole project achieved a RICS 'Silver' Ska Rating. The Ska Rating is an environmental assessment method designed specifically for interior fit-out projects that integrates criteria for certified products and materials such as Environmental Product Declarations and Cradle to Cradle® Certification. Ska also rewards projects that consider the health and wellbeing of the occupants of the building by considering design aspects such as daylight and air quality; the latter is closely linked to the choices of materials and finishes in the fit-out.

An occupant survey carried out by Space Works Consulting, on behalf of Grigoriou Interiors, showed that eight out of ten occupants thought their productivity was higher in this space than elsewhere. Huckletree's first co-working space in Clerkenwell reached its membership capacity after only two months of opening.

Lean Design

A hundred years ago, one of the founders of the engineering company Brown Boveri (now ABB) said that the art of engineering is the elimination of parts.[7]

For every component that is used in a building, there will be associated waste from mining, manufacturing, maintaining and ultimate disposing of the element. Slimming down the design to the bare essentials can create a virtuous circle where both the embodied and the operational impacts of the building can be reduced, as shown in the next case study. The Enterprise Centre case study in Chapter 10 also shows how using lightweight, biological materials and reclaimed components can lead to a leaner design with considerably less resource use.

Unity House, Wakefield

The principle of lean design can provide elegant solutions that integrate the architecture with the building services strategy, whilst saving money and resources. Unity House in Wakefield, Yorkshire, was built in 1867 as the Co-operative headquarters, but it had been derelict for ten years. A local community group, Unity House (Wakefield) Limited, led a campaign to restore this Grade II listed building, but were only able to raise limited funds. When the initial tenders came in 50% over budget, the group turned to LEDA to bring the mechanical and electrical services costs to less than £600,000.[8]

07. DESIGNING-OUT WASTE

An unappealing suspended ceiling had been installed in the main hall sometime in the past. When it was removed, it revealed a decorative wood panelled ceiling and a fully ducted Victorian passive ventilation system leading up to a plinth where an elegant ventilation turret had once stood.

Rather than taking the brute force approach of installing a full mechanical ventilation system complete with associated plant and capital costs, LEDA proposed restoring the original passive ventilation strategy.

Matthew Hill of LEDA was able to source a new ventilation turret that was acceptable to the conservation officers and then reconnected the existing fully ducted system that was originally used to ventilate the hall. This created a passive-assisted extract ventilation system with very little new plant and equipment (see Figure 7.08). As Hill says, lean design 'means forsaking the traditional belt and braces approach to services design for a more holistic strategy'.

The solution bought back to life a redundant building on an affordable budget, restoring the building's original design philosophy, and using fewer resources both in its design and its operation.

Figure 7.08: Unity Hall after restoration.

Conclusion

The design decisions made at the start of a project will have a profound influence on the amount of waste generated. Deciding to refit or refurbish an existing building instead of demolishing and building new is often a difficult path to follow, but there is a demonstrable market for buildings with history. Similarly, using reclaimed components in design requires a more flexible approach where the design is perhaps driven by what is available instead of the original vision. Finding ways to slim down the design to reduce the demands for components has to be done by considering the whole life of the building.

There are opportunities to reduce the demands for complex, resource-intensive building services components and internal fittings by applying excellent design principles, but the structure of the building needs to be robust enough to allow it to be adapted to new uses, as explained in the next chapter.

08. Design for adaptability

Almost no buildings adapt well. They're designed not to adapt; also budgeted and financed not to, constructed not to, administrated not to, maintained not to, regulated and taxed not to, even remodeled not to. But all buildings (except monuments) adapt anyway, however poorly, because the usages in and around them are changing constantly.

Stewart Brand

It is rare for a building to remain occupied by its original owner for long, so even bespoke building designs will have been adapted to meet the needs of new users or will have been pressed into accommodating a new way of working or a major change in technology that brings new demands to the space.

There is a difference between designing a building that is simply flexible and one that can be adapted to new uses. Addis and Schouten differentiate between flexible and adaptable buildings as follows:

- 'Flexible building – a building that has been designed to allow easy rearrangement of its internal fit-out and arrangement to suit the changing needs of occupants.
- Adaptable building – a building that has been designed with thought of how it might be easily altered to prolong its life, for instance by addition or contraction, to suit new uses or patterns of use."[1]

A proportion of the building stock has proved to be adaptable more by accident than by design. Georgian townhouses and Victorian warehouses are good examples of buildings that have been adapted for completely different uses in their time. Wren House in Hatton Garden, London, was built as a church in c. 1670, and then became a charity school c. 1696. The interior was damaged in the Second World War by incendiary bombs and had to be restored. It is now used as an office (see Figure 8.01).

BUILDING REVOLUTIONS

Figure 8.01: Wren House, Hatton Garden.

However, many buildings are not adaptable, suffering from any number of limitations that mean they depreciate in value and are eventually pulled down.

Open Buildings

There are movements and initiatives that have proposed ways to deliberately design buildings to be more flexible and adaptable.

'We should not try to forecast what will happen, but try to make provisions for the unforeseen,'[2] explains John Habraken, who has proposed an approach to design that he calls 'Open Building', a concept that has been widely adopted across the world. One of the aims of Open Building is to provide built environments that last because they can adjust and adapt to change.

Habraken recognised that there are several different, but related, ideas that could be considered when designing the built environment, including some thoughts that relate directly to circular economy thinking:

- The idea of distinct levels of intervention in the built environment, such as those represented by 'support' (the structure) and 'infill' (the fit-out).
- The idea that the interface between technical systems should allow the replacement of one system with another performing the same function, so replacing the building services should not mean removing all the finishes.
- The idea that the built environment is in constant transformation and change must be recognised and understood.
- The idea that the built environment is the product of an ongoing, never-ending design process, in which environment transforms part by part.[3]

As Habraken proposes: '... a strict separation of a long term "primary structure" from a short term "secondary structure" would assure better adaptation to new equipment and changing demands ... over the life time of the building'.[4] Open Building designs aim to separate the 'support' and the 'infill' in the way that the base building is a separate entity to the fit-out. The interfaces between different elements of the building have to be carefully designed to allow each part to be replaced without adversely affecting the other systems. This aligns with the concept of 'building in layers' proposed in Chapter 6. This separation of structure and the infill means that there has to be some level of redundancy in the structure and service cores have to be more generous to allow for future adaptation.

Industrialised, Flexible and Demountable

The Dutch Government applied the Open Buildings philosophy and developed a programme that combines standardisation, customisation and adaptability called 'Industrial, Flexible and Demountable' construction (IFD). The design criteria for an IFD building include:[5]

- integration and independence of disciplines: installation, bearing structure, outer shell and interior finishing
- a completely dry building method: no pouring of concrete, mortar joints, screeds, stuccowork, sealant or polyurethane spray
- perfect modular dimensioning: a great deal of attention to drawings, prototype testing, quality system for drawings, and assembly instructions
- adjustability of all parts: bearing structure (limited), installation (practically unlimited), outer shell (limited and modular), interior finishing (practically unlimited and modular).

Many of these ideas go against the grain of conventional construction techniques, which include a wide range of composite components that are bonded together. For example, in the UK the most popular floor design for multi-storey, non-domestic buildings is the composite steel and concrete deck construction.[6]

Martini Hospital, Groningen

The Martini Hospital exemplifies IFD design and was awarded IFD demonstration status by the Dutch Government (see Figure 8.02). Built in 2008, the SEED Architects' design includes the standardisation of the building skeleton which is made up of uniform building blocks, with the façade panels, system walls fixed furniture and most of the services being completely prefabricated.

Extensions of 2.4 × 7.2m can be added to the outside of the building, allowing the building floor area to be increased by roughly 10% (see Figure 8.03). The partition walls are demountable, allowing spaces to be reconfigured or even to be converted into other uses.[7] The IFD principles are applied at the room level: it is possible to move supply points for electricity, medical gases and water, as well as counters and cabinets.[8]

The internal layout of the hospital can be completely reconfigured, so ward spaces can be changed into outpatient clinics, offices or even apartments, and vice versa.

The Martini Hospital is a successful example of deliberately designing a building to be more flexible, so that it can be reconfigured into a differently designed hospital.

Figure 8.02: Martini Hospital, an adaptable building.

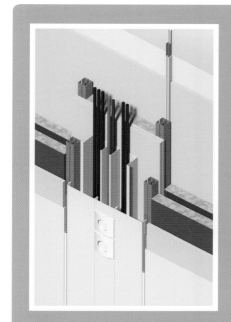 

Figure 8.03: Martini Hospital, adaptable features.

How Buildings Learn

Stewart Brand's research into the way that buildings change and adapt with time is set out in his book *How Buildings Learn: What Happens to Buildings After They're Built*.[9] The book proposes some strategies that may allow building to be more adaptable to change:

- Loose fit structures: spend more money and apply more effort to the structure of the building, less on the finishes and more on adjustment and maintenance.
- Scenario planning: as Brand bluntly puts it: 'All buildings are predictions. All predictions are wrong.'[10] Using scenario planning to determine the alternative potential futures for the building will help to make the building designers think more about how the building could be used. This should help to lessen the chances that the building is so tailored to one use that it is quickly made obsolete.
- Simple plan form: Brand's case study examples show that 'the only configuration of space that grows well and subdivides well and is really efficient to use is the rectangle'. Complex building forms often result in buildings that are harder to change, extend and adapt.

BUILDING REVOLUTIONS

- Shearing layers: as discussed in Chapter 6, applying the idea that buildings have different, independent layers results in a design imperative that 'an adaptive building has to allow slippage between the differently-paced systems of Site, Structure, Skin, Services, Space Plan and Stuff'.

Finally, Brand proposes that architects can mature from being artists of space to become artists of time, considering how buildings might change, and enabling that change.

The Multispace Concept

3DReid created an adaptable building concept based on research into the design parameters for different types of building. The result was a set of design parameters that they called 'Multispace'.

The Multispace research compared different parameters for different types of buildings to see where the overlaps occurred and therefore the opportunities to design for adaptation. The key parameters they examined included:

- storey height
- building proximity
- plan depth
- structural design
- cladding design
- vertical circulation, servicing and core design.

Figure 8.04: Comparison of floor-to-ceiling heights.

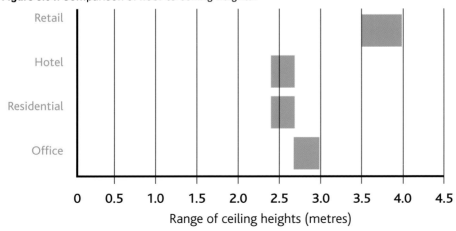

08. DESIGN FOR ADAPTABILITY

For example, a comparison of floor-to-ceiling heights showed that there was some overlap between typical standards for office, residential and hotel (bedrooms), whereas the retail buildings were outside the range (see Figure 8.04).

The research noted that the more generous storey heights allowed a wider range of servicing solutions. The proposed solution was a slim floor construction that allowed the most generous floor-to-ceiling heights without adding too much to the overall height of the building, along with a double-height ground floor that allowed for retail or for reception space.

Figure 8.05 provides a summary of the results of the research showing a range of features that are intended to create a more adaptable building.

Chris Gregory, now of TEC Architecture, conceived Multispace whilst at 3DReid. The two practices prepared a research proposal that examined buildings that had successfully adapted to changes and survived while their neighbours had succumbed to the wrecking ball. One of the buildings examined was Abbey Mill in Bradford on Avon. It was built as a cloth mill in 1875 and is an impressive building located on the river Avon (see Figure 8.06). With the decline of the wool industry, it was eventually sold and was used for rug-making for a few years. During the First World War it was used by the army to billet soldiers. In 1915, the mill was bought by a rubber company

Figure 8.05: Summary of the Multispace concept showing adaptable design features.

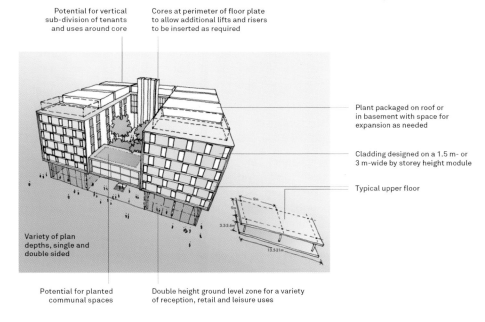

59

Figure 8.06: Abbey Mill, Bradford on Avon, a building that has proven to be very adaptable.

who used the building for storing rubber components and for rubber production. In 1967, the building was converted into offices and restaurant, with the exception of the ground floor. In 1996 it was converted yet again into retirement apartments.[11]

These different uses would normally be considered incompatible with each other, so how has Abbey Mill managed to survive and adapt in the way that it has? TEC Architecture and 3DReid proposed seven insights into why the building had endured:

1. People like it. The building is a simple, elegant structure which dominates the riverside in the centre of town. It is taller than many surrounding buildings but people feel that it forms an important part of their townscape. They want to keep it and therefore they will continue to repair it and find use for it. The power of good architecture should not be underestimated in adaptable design.
2. There is a generous courtyard space associated with the building which acts as a social focus, a service access and a space for expansion. The space outside a building is arguably more important that the space within.
3. The upper floors of the main building are flexible rectangular spaces punctuated only by two rows of columns. They can accommodate almost anything from a production process to a highly partitioned residential care home.

4. The ground floor is deep, closely following the edge of the street and has the potential to take any number of special facilities that would not fit into the standard upper floor plates.
5. The storey height is generous: about 3.2–3.3m. This gives plenty of room for additional services to be added if required. But, more importantly, most of the space can be naturally day-lit and ventilated, which minimises the reliance on services in the first place.
6. The windows are tall, admitting daylight deep into the floor, and regular, at about 3m centres, so there are no 'black spots' which would limit use.
7. The main original staircase is treated as a separate element that serves the main floor spaces without interrupting them. This has allowed a lift shaft to be added and a further fire escape on the far side.

The message is that a simple, robust and elegant structure will endure.[12]

The research came in useful when a volatile London market drove the developers of Grosvenor Dock to hedge their bets and obtain planning consent for three different uses in one of the buildings. 3DReid developed the planning application for a five-storey building on the waterfront that could be residential, office or restaurant/retail use. It was completed in 2007 as residential accommodation (see Figure 8.07).

Figure 8.07: Grosvenor Dock, London, designed for adaptation.

The building includes a generous floor-to-ceiling height. As Chris Gregory notes:

> A loose fit for one building may be just right for another one; the higher ceiling heights required for offices would drive higher ceilings in residential buildings. This would increase costs, but may also add value to the residential buildings, as well as providing the building owners with the ability to adapt the buildings to new uses in the future.[13]

The core is carefully positioned to allow for an open plan office floor plate or flats, whilst avoiding creating dead-end corridors. The design includes soft spots areas within the post-tensioned slab so that new service risers can be added. The soft spots are based on the principle that there should be no more than 30m between the cores to limit the length of the horizontal service runs. Post-tensioned concrete slabs are used as they provide a slim slab depth with no projecting beams and generous spans. This helps to keep the much-prized floor-to-ceiling heights, whilst keeping the overall height of the building within the parameters of the masterplan. This construction also allows space for soft spots, as the post-tensioning tendons are typically between 900 and 1500mm apart.

Conclusion

The research and case studies show that it is possible to design buildings that are more adaptable. Whether they will be successfully adapted depends on economic and social factors as much as technical design-related parameters.

The use of a layered approach allows buildings to be flexed and adapted more readily. In particular, a separation between the primary structure, the façades, the services and the interiors of the building allows the structure to be retained whilst the façade is replaced, or the interiors to be changed into new layouts whilst not being dictated by structural walls in awkward locations. Equally, the building services need to be as accessible as possible and not entangled with the structure or encased within walls, floors or behind finishes.

Other factors that would help to make the building more open to adaptation are:

- over-engineering the structure and foundations to accommodate changes in use and the potential expansion of the buildings
- exploring different structural solutions that provide clear floor plates and options to add or remove parts of the floor plate or to provide penetrations through the floors for different servicing strategies
- using a simple plan form to allow reconfiguration of internal spaces and expansion
- positioning of cores and generous sizing of risers to allow for future changes in use

08. DESIGN FOR ADAPTABILITY

- scenario modelling to show how the proposed buildings can be adapted to different uses – this will help to inform the optimum floor-to-ceiling heights, service zones, floor plates and structural grids that can accommodate different uses
- generous floor-to-ceiling heights with abundant levels of daylight (a feature of many buildings that have been through several adaptations) and the provision of external space
- applying the principles of design for disassembly (see Chapter 9), which should make the building more adaptable as the building components can be separated more easily and different elements will be more accessible for repair or replacement.

Building-in the ability for interiors and local services to be adaptable to new uses is potentially wasteful as it is likely that these elements would be stripped out and replaced if the building is adapted to a new use. Designing-in some degree of flexibility to allow internal partitions to be relocated and local services to be extended or augmented may be appropriate, depending on the building use and the lifespan of the components.

Designing simple, robust and elegant buildings that are valued by people helps to make sure that buildings endure. Alternatively, buildings can be designed as modular structures that can be demounted and reconfigured, as explored in the next chapter.

09. Design for disassembly and reuse

At its core, a circular economy aims to design out waste. Waste does not exist: products are designed and optimized for a cycle of disassembly and reuse.

World Economic Forum

Being able to extract components from buildings intact is an important part of applying the circular economy. Components that have reached their end-of-life or need repair have to be easily accessible, rather than buried in other materials. Elements of a building that are to be removed as part of a refurbishment, refit or demolition have to be recovered without damage so that they can be reused, remanufactured or recycled.

Materials that are toxic have to be isolated and returned to industrial processes. Biological materials have to be separated and uncontaminated so that they can be composted or used to generate biogas.

To achieve these aims, buildings have to be designed with a thought for what happens to their constituent parts at the end-of-life.

Designers will often consider plant replacement strategies as part of the design and will work up drawings showing how large components, such as chillers, can be removed and replaced. If the brief demands, designers may also prepare strategies showing how cladding panels can be replaced or internal partitions can be relocated. It is conceivable, though rarely done, to have a strategy for reclaiming components and materials at end-of-life, and to design to enable disassembly of the building.

This chapter summarises the principles for deconstruction and includes some case studies showing how a few designers are thinking about the end-of-life of buildings and how they can be deconstructed and used again. Chapter 12 sets out some different business models that turn buildings into materials banks for future generations.

Principles of Designing for Deconstruction

There is a wealth of research about designing buildings for deconstruction (or disassembly) and various guides that set out principles to be followed. Notably, a guide by CIRIA[1] sets out principles (Table 9.01)[2] and design guidance, by building element, on designing for deconstruction.

A survey of demolition contractors[3] showed that some of these principles are considered more beneficial to deconstruction than others, namely:

- having mechanical and reversible (e.g. not chemical) connections
- ease of access to connections
- independent and easily separable elements of the building, e.g. structure, envelope, services and internal finishes; and
- no resins, adhesives or coatings on the elements.

Table 9.01: Design principles for deconstruction for different end-of-life outcomes.

DESIGN PRINCIPLE	COMPONENT REUSE	COMPONENT MANUFACTURE	MATERIAL RECYCLING
Provide identification of materials and components	✓	✓	✓
Provide guidance for deconstruction	✓	✓	✓
Design for simultaneous, parallel disassembly and deconstruction	✓	✓	✗
Design for deconstruction using common tools and equipment rather than bespoke tools	✓	✓	✗
Minimise the number of different types of components	✓	✓	✗
Mechanical in preference to chemical connections	✓	✓	✗
Consider using modular construction	✓	✓	✗
Provide good access for deconstruction, especially connections	✓	✓	✗
Design components sized to suit appropriate means of handling	✓	✓	✗
Provide adequate tolerances for assembly and deconstruction	✗	✓	✗
Design connectors, fixings and components for repeated use	✗	✗	✗
Consider using standard grids	✗	✗	✗
Use the minimum number of different types of connectors	✗	✓	✗
Use the minimum number of interfaces and connectors	✓	✓	✗
Consider the use of prefabrication	✓	✓	✗
Minimise the number of different types of materials	✓	✓	✓
Use alternatives to toxic and hazardous materials	✓	✓	✓
Make inseparable sub-assemblies from the same material	✓	✓	✓
Eliminate the use of secondary finishes to materials	✓	✓	✓
	✓	Very important	
	✗	Less important	

09. DESIGN FOR DISASSEMBLY AND REUSE

The survey also identified the need for information about the building, including full as-built drawings and a deconstruction plan to help the demolition contractor to understand how to disassemble the building.

Doctoral research by Paola Sassi into 'closed loop material cycle construction' proposes a set of technical criteria that can be used to determine whether a material or component can be disassembled.[4] This research is based on an extensive review of literature and analysis of the essential criteria. Table 9.02 shows a summary of the results and examples of compliant and non-compliant components.

Table 9.02 can be used by designers to test out whether their proposals can be deconstructed at end-of-life.

The research by Sassi proposes the following:

- Access to building components and to fixings is essential. Embedding elements within other elements makes it impossible to separate them without damaging one or both of the components. Similarly, fixings need to be easily accessible and not require excessive force to undo them. For example, a fixing cast in concrete would be inaccessible.
- Loose connections and friction fittings are inherently demountable, with mechanical fixings having a relatively high demountability and chemical fixings being the least demountable.

Table 9.02: Summary of technical criteria for design for deconstruction.

PROCESS	REQUIREMENT	COMPLIANT EXAMPLE	NON-COMPLIANT EXAMPLE
Ability to access	All components are readily accessible and removable	Door	Service duct embedded in concrete
Accessibility of fixings	All interfaces/connection points, fixings are identifiable and accessible	Screw-fixed timber cladding	Plasterboard fixings under full coat of plaster
Types of fixings	Fixings have: • the ability to be removed (mechanically, or with solvents) from the element or alternatively integrated within the recycling process • the ability to ensure a high percentage of recovery of the material • the ability to ensure a high quality of the material recycled without contaminants • the ability to remain operational long term	• Water-based / soft adhesive • Screw/bolt	• Insoluble adhesive • Rivet
Durability of fixings	Design joints will remain operational and do not compromise removal of element over time	Stainless steel fixing	Steel fixing that may rust
Information	Sufficient information is provided OR no information is required to enable dismantling	Brick wall	Proprietary temporary building

67

- A fixing should not contaminate the elements. For example, a rusted metal fixing in timber may break off and remain embedded in the wood. On the other hand, organic polymer adhesives can be burned off metal sections when being melted down for recycling without affecting the quality of the new steel.
- Specifying mechanical fixings is not sufficient to enable disassembly; the fixings also have to be durable. For example, using screws that corrode quickly will not facilitate disassembly.

The CIRIA report notes that: 'The value of a building element at deconstruction will reflect its condition at that time, the work needed to refurbish it for later reuse and the level of performance that it might be able to achieve.' Therefore, components or elements need to be easily cleaned, maintained and serviced. For reuse after deconstruction, the elements have to be removed from the building with as little damage as possible.

Difficult Demolition Wastes

Research by BRE in collaboration with the National Federation of Demolition Contractors identifies materials that are currently being used that are difficult to reclaim or recycle at the demolition stage.[5] Table 9.03 shows some of the products identified in the report that are currently entering the waste stream, along with the recovery issues and potential opportunities for recovery.

Table 9.04 shows some of the products that the BRE report suggests will be entering the waste stream in the future.

The BRE report shows that elements that are bonded together or contain hazardous materials create problems at end-of-life, based on current demolition practices. Table 9.03 illustrates the legacy left by the previous generation of buildings and the issues that were not considered when the buildings were designed and constructed. Table 9.04 provides some examples of components that could create new problems when the next generation of buildings is demolished. The list includes components that are designed to improve the efficiency of the building in operation. Even elements that are jointed with mechanical fixings, such as the composite floor cassettes, are noted in the report as being difficult to handle using mechanical demolition.

This is corroborated by research by the University of Cambridge, which shows that even structural steel beams with bolted connections that could be reclaimed are typically cut into sections using hydraulic shears.[6] The additional time taken to unbolt the complex joints will not provide sufficient return to justify the extra labour required. The research goes on to propose that the use of novel jointing techniques such as Quicon, ATLSS and ConXtech could reduce the time taken to facilitate greater reuse in the future.

Table 9.03: Examples of difficult demolition wastes currently entering the waste stream. Summarised from BRE, *Dealing with Difficult Demolition Wastes*

PRODUCT	RECOVER ISSUES	RECOVERY OPPORTUNITIES
Aerated concrete blocks	• Bonded with mortar • Not suitable for recycling into concrete aggregate	• Can be used as a substrate for green roofs
Asphalt roofing	• Classed as hazardous waste • Bonded to roof structure	
Extruded polystyrene foam (XPS)*	• Older board may contain ozone-depleting substances (hazardous waste) • Floor insulation is encased in concrete	
Laminated wood	• Not readily recyclable	• Potential to be composted. • Energy recovery
Metal insulated panels	• Metal insulated panels are typically recycled for the metal. • Insulation is stripped out, and typically is sent to landfill	• Post-2004 panels taken out of buildings can be reused (as these will not contain ozone-depleting substances)
Phenolic foam boards	• Mostly landfilled	• Opportunity to incinerate
Phenolic pipe sections	• Usually sent for disposal as hazardous waste at a landfill	• Opportunity exists to incinerate • Potential for mechanical recycling
Polymer composites	• Typically landfilled in the UK	• Several recycling options have been developed, including the reintroduction of ground Fibre Reinforced Polymer waste into the production process
Polyurethane rigid foam (PUR)	• Boards containing ozone-depleting substances (e.g. pre-2004) need to be disposed of as hazardous waste	• New approaches to reuse and recycling
Wood-based board	• Not readily recyclable due to additives and adhesives	• New technology is being developed to recycle these products, using microwaves and moisture to burst the bonds in the board without damaging the fibres

Table 9.04: Examples of difficult demolition wastes in the future. Summarised from BRE, *Dealing with Difficult Demolition Wastes*

PRODUCT	RECOVERY ISSUES
Brick slips/brick tile system	Bonded to composite insulation panels, so difficult to separate
Closed-panel timber frame: phenolic foam applied in factory	Time-consuming to separate, mineral wool and phenolic foams generally end up in landfill
Concrete with macro fibres (steel fibres, plastic fibres)	Not known whether this material can be recycled back into aggregates
Floor cassettes (composite)	Difficult to segregate materials on site
Insulated plasterboard	Difficult to segregate on site
Insulating concrete formwork	Difficult to segregate on site, risking contamination of recyclable material
Phase-change materials	It is not currently known how PCMs would affect recycling
Roof cassettes	Slow to dismantle and segregate materials

Flooring systems are one of the most intractable problems when designing for deconstruction, as they are often composite systems using steel and concrete to span large distances.

There are some examples of buildings that have been designed for disassembly. The Prologis Distribution Facilities Building near Heathrow was fabricated with a two-bay portal frame and all the steel members are stamped with the section size and grade to allow them to be identified and reused. Approximately 80% of the portal frame structure is designed to be reusable, along with 95% of the floor beams and 100% of the galvanised steel components (SteelConstruction.info). Portal frame buildings are one of the few types of construction that are dismantled and reused, as there is a market for second-hand portal frames for use in the agricultural industry.

Some designers have taken the idea of designing for deconstruction to heart and have created buildings that can be disassembled and recycled with relative ease, as shown by the first case study from Werner Sobek. The second case study shows how Suitebox has gone further by developing temporary buildings that are designed entirely on the basis that they will be disassembled and the components reused to form other buildings. The last case study explains how XX Architecten has created a 'building' from modular components that is now in its third incarnation, having had two previous lives on two different sites.

Werner Sobek

In Germany, Werner Sobek has long been an advocate of environmentally sustainable buildings and has designed prototype houses that are highly energy efficient and that consider the end-of-life recyclability of the components and materials.

His aim is to 'build houses that use no energy, produce no emissions and that are completely recyclable – in the sense of a cradle-to-cradle design, i.e. not downcycling but recycling along the lines of "My house is going to become a Porsche" or "I used to be a can".'

The striking F87 Efficiency House Plus, in Berlin, was designed as a demonstration project that would stand for two to three years before the materials and components were reclaimed (see Figure 9.01). The aim was to show that a building could be designed that could generate more energy on site than is needed to operate it, as well as creating a building that could be completely disassembled at end-of-life.

Wherever possible, the materials are selected as either biological materials that could be composted or technical materials that could be recycled. Approximately 20 different types of materials are used and defined as recyclable at end-of-life including:

- cellulose insulation
- recycled rubber as protective matting

09. DESIGN FOR DISASSEMBLY AND REUSE

Figure 9.01: F87 (Efficiency House Plus).

- wooden bearers to form the structure of the roof and upper floors
- hemp insulation, and
- cork board.

Then, for the façade, timber strips are used along with aluminium slats, triple glazed windows and solar panels that can be reclaimed and reused (see Figures 9.02, 9.03 and 9.04).

The majority of connections are made by means of easily separable screw, click and clamp connectors. The floor, wall and roof constructions are designed in separate layers and adhesives are avoided, wherever possible. The wall and floor coverings are mounted without adhesives to allow reconfiguration and recycling. The wood cladding for the picture window is larch, which weathers naturally and requires no treatment. Similarly, the floor of the terrace and the picture window is made of solid oak, meaning that no chemical protection is required.

A recycling manual was prepared for the building that details the various materials used and the potential for reclamation or recycling.

BUILDING REVOLUTIONS

Figure 9.02: F87 roof structure.

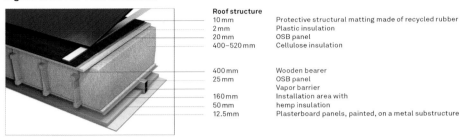

Roof structure	
10 mm	Protective structural matting made of recycled rubber
2 mm	Plastic insulation
20 mm	OSB panel
400–520 mm	Cellulose insulation
400 mm	Wooden bearer
25 mm	OSB panel
	Vapor barrier
160 mm	Installation area with
50 mm	hemp insulation
12.5 mm	Plasterboard panels, painted, on a metal substructure

Figure 9.03: F87 upper floor construction.

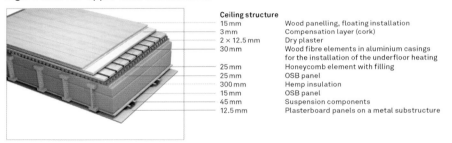

Ceiling structure	
15 mm	Wood panelling, floating installation
3 mm	Compensation layer (cork)
2 × 12.5 mm	Dry plaster
30 mm	Wood fibre elements in aluminium casings for the installation of the underfloor heating
25 mm	Honeycomb element with filling
25 mm	OSB panel
300 mm	Hemp insulation
15 mm	OSB panel
45 mm	Suspension components
12.5 mm	Plasterboard panels on a metal substructure

Figure 9.04: F87 wall construction.

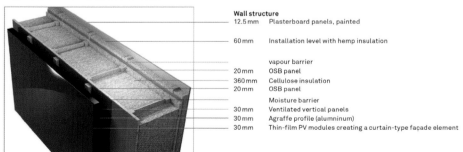

Wall structure	
12.5 mm	Plasterboard panels, painted
60 mm	Installation level with hemp insulation
	vapour barrier
20 mm	OSB panel
360 mm	Cellulose insulation
20 mm	OSB panel
	Moisture barrier
30 mm	Ventilated vertical panels
30 mm	Agraffe profile (alumninum)
30 mm	Thin-film PV modules creating a curtain-type façade element

09. DESIGN FOR DISASSEMBLY AND REUSE

This project shows that highly efficient houses with complex details to reduce heat loss can be designed for deconstruction if the philosophy is considered from the very start of the project.

It was intended that the building would be dismantled after three years (it was completed in 2011), but it has been so popular with visitors that the Federal Government has extended the building's lifetime. At the time of writing F87 is still standing.

Suitebox, London

Who better to ask about designing demountable, reusable buildings than the people whose business depends on it? ES Global, based in London, provides temporary, demountable structures for events, ranging from gantries and staging through to pavilion buildings. The trusses used to form these structures are often 15-20 years old and are shipped all around the world.

Suitebox, a subsidiary of ES Global, also based in London, was set up on the back of the experience and knowledge gained from designing these rapidly deployable, demountable and reusable structures. The aim is to marry the experience of the events industry with the technology of conventional buildings to develop buildings that can be rapidly erected, relocated or reused.

Jon Baker, the founder of Suitebox, has recognised that in some sectors there is a latent demand for buildings that can respond and flex quickly to change. Distribution companies are a good example. Customer demand for faster deliveries has made the centralised, one million square foot distribution warehouses unwieldy and unresponsive. Instead, companies are having to provide more, smaller warehouses that are located closer to customers. For rapidly changing portfolios, there is argument for having a kit of parts that can be deployed in one location and then redeployed at another as the market changes.

The retail sector has similar pressures where market demands mean that their assets have a short life and need to be reconfigured or relocated regularly to keep pace with changing consumer habits. Some retailers are already deploying temporary structures to keep customers happy while a store is refurbished or rebuilt.

There is a potential application in the education sector, with schools often needing temporary classrooms to accommodate growth and the need to flex and change. Local authorities in the UK are also under pressure to provide additional housing. There is the potential to provide homes on brownfield sites that can be exploited in the short term, but may be earmarked for future development or have difficult ground conditions.

Two examples of Suitebox's demountable buildings are discussed on the next two pages.

BUILDING REVOLUTIONS

3i Waterloo Road, London

3i Ltd was looking for somewhere to accommodate its staff while looking for offices elsewhere. Urban Salon Architects worked with Suitebox to develop a temporary building on a site adjacent to their existing office (see Figure 9.05). The building on Waterloo Road, London, employs a steel frame with a lightweight cladding and roofing system that is removable and can be relocated. The floors are made from a demountable steel deck system.

It has a high level of internal fit-out that gives the building the feeling of solidity associated with conventional construction (see Figure 9.06). The building was erected in only 20 weeks and is fully demountable. The ground floor is let as retail and the remaining floors are occupied as offices.

The building was completed in 2004 and has been so popular with occupants that it is still standing at the time of writing.

Figure 9.05: 3i Waterloo Road.

Figure 9.06: 3i Waterloo Road internal fit-out.

Chobham Manor, London

Suitebox has been developing its approach to demountable construction by refining its kit of parts and has tested out new designs. A part of this development work is evident in the marketing suite for Taylor Wimpey (completed in 2014) on the London 2012 Olympic Park site. The Chobham Manor site is contaminated and will be remediated and redeveloped as part of the Olympic legacy. Taylor Wimpey wanted a temporary building with the look and feel of a permanent building to provide office accommodation and a show home for the development.

Octink, in conjunction with Suitebox, designed a three-storey temporary building with no foundations, to avoid the need to pile or excavate the contaminated ground. The foundations are pads laid onto compacted ground and the steel frame is pinned together, rather than bolted, to ease assembly and disassembly (see Figure 9.07). The cladding is of wood panels with larch. The ground and upper floors are cassette units made up of 18 mm phenolic-faced plywood, multi-foil insulation and steel stretchers. Some of the floors need gyproc fire-resistant panels for fire protection. The cassettes are bolted down using 'Speedthread' fixings consisting of a bolt and an oversized washer that allows the corners of the panels to be fixed down together and easily demounted. As Jon Baker says: 'The first thing we look at is how it comes apart, before we think about how we're going to build it.'

Figure 9.07: Chobham Manor Marketing Suite showing structure and foundation pads.

Figure 9.08: Chobham Manor Marketing Suite, completed building.

Suitebox is aiming to further refine the design for future projects so that more of the construction can be reused. In particular, the current roof design is a composite roof membrane which will not be reused, so Suitebox is developing a cassette system that can be demounted and redeployed. The larch cladding was painted, but on future projects the aim would be to leave the wood untreated as larch will weather naturally and does not require finishes.

The intention is that the building will be in place for five or six years after which time it can be disassembled and completely removed from the site, leaving no foundations or floor slabs that would hinder the redevelopment of the site. Suitebox has guaranteed buyback on some of the elements.

These case studies show that it is possible to design buildings that are fully demountable without any loss of quality and that there are applications for this technology. The next case study shows how modular, demountable buildings can be reincarnated.

XX Architecten, Rotterdam

XX Architecten has an impressive track record in designing demountable buildings for a limited lifespan that can be redeployed for different purposes or fully disassembled, allowing the materials and components to return to the biosphere or to be reused.

Villa Camera is a rare example of an Industrial, Flexible and Demountable (IFD) building (see Chapter 8 for an explanation of IFD) whose components have actually been demounted and reused as three different buildings.

Villa Zebra, Rotterdam

In its first life, the building was the Children's Hall of Art in Rotterdam (see Fig 9.09). Completed in 2001, the aim of Villa Zebra was to create a cultural workspace for children that included exhibitions, cooking areas, a children's café, a theatre and studio spaces. The building was designed to have a five-year life with the ability to be extended or decreased in size, depending on the success of the venture.

XX Architecten designed a flexible, demountable building from modular steel units (6 × 3 × 3m) with mechanical connections between the units and prefabricated cladding panels.

After five years on the site, it was dismantled and the components were stored.

Figure 9.09: Children's Hall of Art in Rotterdam.

Villa Nutcracker, Hoogvliet

Two years later the same elements were used in a different combination to build temporary accommodation for a school in Hoogvliet. Villa Nutcracker was intended to last for three to five years until a new school building was constructed. This time, the building was made 3m wider by adding additional units to allow the classrooms to be interconnected. Villa Nutcracker contained nine classrooms, common areas and a library.

Villa Camera, Hilversum

Then, in a remarkable third reincarnation, Villa Nutcracker was dismantled and rebuilt into three new buildings. A section of the lower part of the building has become a small school building, whilst another section has been added to an existing harbour building for Co-dart, a school for artists in Rotterdam. Part of the building has been reassembled as Villa Camera (see Figure 9.10). Designed for Facility House Broadcast Group, the building is located in the Media Park in Hilversum and is used to store camera equipment for hire along with associated office space. Opened in 2013, it has been nominated for an architecture reward because of the combination of expressive architecture and nature.

09. DESIGN FOR DISASSEMBLY AND REUSE

Figure 9.10: Villa Camera Facility House, Media Park, Hilversum.

As Jouke Post says: 'It's almost a new type of building: a nice building made of boring boxes.'[7]

Conclusion

Modular, demountable buildings can be deployed temporarily on sites that are earmarked for redevelopment.

Designing for deconstruction is an essential piece of the circular economy puzzle. It enables buildings to be deconstructed rather than demolished, and when done correctly, it should make the process safer and more efficient. It allows components and materials to be reclaimed intact during renovation or demolition. If this is combined with a new infrastructure, mechanisms to make buildings into materials banks and inventories that allow materials to be pre-sold before the building is even demolished, then there is the potential to close the loop. These ideas are explored in Chapter 11.

As well as creating mechanisms to encourage salvage and reuse, the materials and components have to be selected to enable reclamation, remanufacture, recycling or composting. This is discussed in the next chapter.

10. Selecting materials and products

Buildings are gold mines of materials just waiting to be harvested.

Ellen MacArthur Foundation

In a circular economy, each material in the building and its components has to be declared and defined, the composition has to be as pure as possible and it has to separable from other materials. These requisites ensure that buildings can become materials banks (see Chapter 11) where materials are effectively stored for future use, rather than consumed and lost.

One of the defining principles of a circular economy is the distinct split between 'biological materials' and 'technical materials', as explained in Chapter 1. Biological materials, such as wood and sand, are used in products that are free of contaminants and toxics, so they can be returned to the biosphere at end of life. Technical materials, such as metals and plastics are retained within industrial loops that ensure they are not lost to the economy or returned to the environment. These technical components should be reused as far as possible and the constituent materials recycled as a last resort.

Matching Lifetime to Material Selection

The idea of 'building in layers' (proposed in Chapter 6) aims to differentiate between components with a shorter lifespan (e.g. carpets) and those with a longer lifespan (building structure). The likely lifespan of each component should be carefully considered and the materials selected accordingly. The lifespan of internal fixtures and fittings is often over-estimated and designed for a long lifespan that is not achieved, leading to significant waste. Components that are likely to have a shorter expected lifespan can either be made from biological materials that can be returned to the biosphere or designed to be readily returned to the manufacturer for reuse, remanufacture or recycling. Components with a longer lifespan, such as the structure and fabric, should be designed to be durable and resilient, whilst ensuring that they can be maintained, upgraded or disassembled, as required.

'Project XX', Delft

Project XX is an experimental office building design in Delft, Netherlands. The aim was to design a building with a limited lifetime, on the basis that office buildings often undergo major refurbishment every 20 years. So rather than designing a building that is supposed to last for 60 or 100 years, why not design one that just lasts for 20 years? The name 'Project XX' represents the design life of the building in Roman numerals.

Each of the building elements and materials were classified according to their lifetime and their ability for reuse. The main elements of the building were chosen to last for the required 20 years and then the following criteria were applied:

- simple to reclaim as uncontaminated raw materials, such as sand or untreated timber
- reusable without any alteration, in general applications
- reusable with minor alterations, in specific applications
- fully separable and recyclable.

Jouke Post of XX Architecten says: 'Materials should be matched to the expected lifespan of a building. That means that components with differing lifespans should be mounted so that they can be dismantled separately.'

The building uses some simple techniques to increase the potential recovery of materials and components:

- the façade is independent of the structural frame
- all the connections (e.g. steel plates, pins and bolts) can be dismantled and avoid the use of glue, putty or sealant
- the insulation is not bonded to the prefabricated ground slabs or the roof membrane
- the internal timber cladding panels are left unpainted, and
- the carpet is not glued down.

10. SELECTING MATERIALS AND PRODUCTS

Figure 10.01: Project XX, Delft, showing interior.

Dry methods of connection are used between different elements, such as:

- the façade frame and the glass panels
- the steel plate connections between timber columns and primary beams, and
- the connections between floor panels and beams.

The building also uses biological materials that can be returned to the biosphere at end-of-life, so the ventilation ducts are made from cardboard, the structural frame is laminated timber (see Figure 10.01), and a sand fill is used in the first floor to provide acoustic insulation.

The building was constructed in 1996 and has proved so popular with the occupants that it remains standing at the time of writing.

Biological Materials

Biological materials are those that can be returned to the biosphere and allowed to biodegrade. The circular economy model proposes that the use of biological materials can be 'cascaded' through various uses, rather than used just the once. An example of cascading might be where solid timber is used in a building and then chipped for use in panel products.

Biodegradable materials are those that can be broken down by microorganisms in the biosphere. Sassi proposes that biodegradable materials can be grouped into four categories:[1]

1. Natural materials that can be used following minimal processing (e.g. timber, bamboo, cork, hemp).
2. Natural materials bonded with a resin or mesh (e.g. clay, hemp and straw mixtures for external walls, strawboard for internal partitions, jute carpet backing, linseed oil and natural resin to make linoleum).
3. Natural compounds used in manufacturing products including adhesives and other polymers (e.g. natural protein to manufacture biodegradable plastics).
4. Biodegradable synthetic materials (biodegradable plastics).

Materials in the first category require only to be shredded to allow them to be composted. For materials in the second category, the aim would be to ensure that biodegradable material is not bonded with non-biodegradable materials (e.g. natural fibres used in concrete or cement products or chemical, non-biological bonding agents), as this will inhibit its ability to be returned to the biosphere. Non-toxic bonding agents can be used in minimum amounts, but this would restrict the potential uses for the resulting compost.[2]

Natural biodegradable plastics can be made from polymers such as cellulose, starch, protein and sugar molasses extracted from plants.[3] Biodegradable synthetic plastics have been developed and have some limited use for disposable packaging and plastic bags.

Substituting biodegradable materials for technical materials is one way to reduce the end-of-life impact of components, particularly those with a short lifespan or those that currently have poor reclaim or recycling rates. Examples are shown in the following case studies.

Adaptavate – a Biocomposite Plasterboard

Fit-outs and refurbishments often involve stripping out plasterboard partitioning and drylining. All too often, the plasterboard is scrapped and sent to specialist landfills, despite it being technically and commercially viable to recycle the gypsum into new plasterboard. There are alternative wall boards on the market, but there are very few truly natural products that can be composted at end-of-life.

Thomas Robinson researched biocomposites during his MSc at the Centre for Alternative Technology and his idea for a natural wall board product was selected for the UK Climate-KIC Accelerator Programme based at Imperial College London.

10. SELECTING MATERIALS AND PRODUCTS

Figure 10.02: Breathaboard biocomposite plasterboard alternative.

This allowed him to raise enough funds to set up Adaptavate and to start the product development of a biodegradable board called 'Breathaboard' (see Figure 10.02). The board is made from a biocomposite with the aim of providing the following beneficial properties:

- Hydroscopic, meaning that it will absorb and release moisture from the air. This will help to reduce mould growth and the impacts this can have on the health of occupants. It should also help to reduce the impact of moisture on the building fabric.
- Entirely compostable at end-of-life, meaning that offcuts or the whole board can be returned to the biosphere.
- Lighter than other wall boards, making it easier to lift and handle.
- The binder has been proven in other applications to absorb VOCs. Robinson is hoping to be able to demonstrate that his wall board will also absorb VOCs from the internal environment.

Plasterboard is an example of a ubiquitous building component currently made from a technical material that is often not returned back to an industrial process. Products like Breathaboard represent an opportunity to substitute a biological material for a technical one, providing multiple benefits in the process.

GatorDuct – Cardboard Ductwork

Building services typically have a shorter life than the buildings that house them. Systems and components that service floors and occupied areas are often stripped out as part of a refit, particularly in office and retail environments. Perhaps in these situations, the local services could be considered as consumables. So ductwork could be made of readily recyclable materials instead of using durable, engineered metals and plastics. There is ductwork on the market made from fabric and even from cardboard.

Tri-wall cardboard ductwork has been developed by GatorDuct, which has a coating made from a water-based solution with a water dispersal polymer, fire retardant minerals and a final hydrophobic finish (see Figure 10.03). The coating can be recycled with other water- and oil-based printed paper. The cardboard is high strength and can be used by recyclers for other substantial cardboard products. Cardboard ductwork requires less insulation than steel ductwork as it has some insulating properties. It is also considerably lighter, making it easier to handle and install.[4]

Figure 10.03: Cardboard ductwork.

10. SELECTING MATERIALS AND PRODUCTS

Architype and The Enterprise Centre

Biological materials can be used to create lighter construction that uses fewer resources, as well as providing materials that can be returned to the biosphere.

Architype's desire for simplified designs can be traced back to its roots in self-build housing. Designing for lay builders means everything has to be as simple and pared back as possible. The practice has carried this philosophy through its projects and has designed buildings that use less resources through a combination of lean design and careful materials selection.

The self-build influences are evident in the lightweight timber frame buildings that are a common theme of Architype's designs. The Coed-y-Brenin Visitor Centre in Wales exemplifies the lean, low impact approach. Its lightweight structure uses locally sourced Sitka spruce, which is turned into structural-grade timber by using short lengths of softwood held together with hardwood dowels that swell and lock the planks together. The cladding is formed from charred local wood that means it needs no other finish. The jointing methods and lack of finishes mean that there is no need for glues, adhesives or paints.

As part of the lean design philosophy, Architype champions the Passivhaus standard, an approach that aims to create comfortable buildings with dramatically lower energy demand. The standard combines high levels of insulation and airtightness with mechanical ventilation and heat recovery to reclaim any escaping heat. Cooling demand is reduced by using shading, pre-cooling the air and by using natural ventilation and night cooling strategies. Architype finds that applying these techniques on their buildings allows them to slim down the amount of heating and cooling kit to a fraction of their typical sizes.

All this experience came in useful when the University of East Anglia released its demanding brief for The Enterprise Centre (see Figure 10.04). The university wanted a new gateway building for its campus that could accommodate an innovation lab, a 300-seat lecture theatre, flexible workspace, teaching and learning facilities, and incubator units for new start-ups. And it wanted it to reduce its embodied carbon and operational carbon emissions by using innovative construction techniques and designing to the Passivhaus standard.

Focusing only on reducing embodied carbon does not necessarily fit into the circular economy ideal, as it can drive designers to substitute highly recyclable (and recycled) materials, such as metals, with materials with lower embodied carbon – for example thermoset plastics, which are difficult to recycle. Also, focusing on embodied carbon does not consider the other impacts associated with winning and processing the raw materials, such as its scarcity or the impact on biodiversity of mining or drilling operations.

Figure 10.04: The Enterprise Centre, University of East Anglia.

Figure 10.05: The Enterprise Centre at UEA, internal view.

10. SELECTING MATERIALS AND PRODUCTS

For The Enterprise Centre, Architype was careful to consider all the environmental impacts over a 100-year life by consciously selecting biological materials. The designers worked closely with the contractor, Morgan Sindall, to select materials and components that were locally sourced, natural and biodegradable. This helped them to avoid the impacts of mining and transporting materials from abroad, whilst stimulating the local economy.

The most striking and innovative feature is the use of thatch to clad the building, which draws on the local vernacular and gives it a novel twist. The design uses prefabricated thatch panels constructed as cassettes that are slotted into place, providing additional insulation as well as acting as a rainscreen. The thatch is simple to repair or replace and the straw can be composted at end-of-life. The clerestory rooflights are also thatched using traditional Norfolk reed over the pitched roof.

The timber frame building uses studwork mainly sourced from the local Thetford Forest and the structural timber columns under the main entrance canopy are made from local trees turned in Suffolk. Cellulose insulation made from recycled newspaper is used as both the thermal insulation to the walls (Warmcell) and as acoustic insulation and the ceiling finish (SonoSpray).

The designers were able to reap the rewards from selecting a lightweight frame and cladding system by radically reducing the amount of foundations required. The lean design of the foundations uses a concrete floor slab constructed with aggregate from a hospital that was demolished nearby, and it uses the maximum 70% GGBS (Ground Granulated Blast Slab) cement replacement. The results are seriously impressive: the foundations and floor build-up are a tenth of the embodied carbon of a conventional design solution and the durable diamond ground floor finish means that replacement cycles of the floor finish are all but eradicated.

Biological materials are used for the interior finishes, including Clayboard cladding with an earth finish in the atria. Additional finishes were avoided wherever possible, as tiles and carpets increase the embodied carbon and paints, grouts and adhesives would inhibit the deconstruction of the building. The use of natural, breathable materials provides a healthier internal environment for the occupants (see Figure 10.05).

The team also salvaged components from other buildings to reduce the demands for virgin materials. The front reception desk is an unused desk from The Sainsbury Centre for the Visual Arts (located on the UEA campus) and the timber cladding to the west elevation uses reclaimed Iroko timber from the original UEA laboratory desks, which were taken out of storage and given a new life by simply cutting and planing them to size.

The passive design principles mean that the building services and distribution systems are far smaller than for a typical building with domestic-sized radiators, while minimal ductwork lengths use transfer ventilation grills as part of the heat recovery strategy.

The result is a lean building, with an overall embodied carbon calculated to be 440/kg/CO$_2$/sqm across the 100-year lifecycle. This equates to around a quarter of the lifetime emissions of a conventionally constructed university building of equal size and scale. Comprised of 80% biological materials, the majority of the building can be returned to the biosphere at the end of its life.

Technical Materials

Technical materials are those that are retained within industrial cycles. When used in the structure or fabric of the building, technical components have to be designed so that they can be salvaged for reuse, or re-engineering, with recycling as the last resort. When technical materials are used in plant and equipment, the components have to be designed to be accessible for repair or replacement by 'building in layers' (as discussed in Chapter 6), and can also be designed for upgrade and remanufacture by being modular. In addition, some elements can be leased from manufacturers rather than purchased (see Chapter 12).

The vast array of technical materials that are available to designers, manufacturers and contractors allows the development of incredibly complex components using materials with hundreds of different polymers and alloys. The components are made from materials that are typically not catalogued and can be combined irreversibly with other, very different, materials into what McDonough and Braungart call 'monstrous hybrids'.[5]

This vast array of materials makes recycling difficult. Even metals that are valuable enough to be reclaimed and recycled from buildings can be difficult to recycle at end-of-life. Allwood and Cullen note in *Sustainable Materials: With Both Eyes Open* that it is not possible to remove the impurities from aluminium, so most recycled aluminium is used as casting alloys rather than wrought aluminium,[6] which has a lower value.

Many plastics can be segregated for recycling, but only if they are separated from other types. There are two main types of plastics: thermoplastics that can be melted and reformed and thermosets that cure irreversibly and so cannot be recycled into similar grade products. Building components use a wide range of plastics, including thermoset plastics such as polyisocyanurate (PIR) and phenolic foam for components such as insulation. These materials are typically sent to landfill or incinerated rather than being recycled.[7] Thermoset plastics can be downcycled into a lower grade product by crumbing and remoulding with new material.

Some manufacturers have recognised the problem with a wide range of materials and have worked to reduce them: Hewlett Packard has been steadily reducing the number of plastics used across its product range from 200 to six to make it easier to recycle them at end-of-life.[8]

Biomimicry and 3D Printing

According to Michael Pawlyn in his book *Biomimicry in Architecture*, 'Nature uses a very limited subset of the periodic table whereas we use virtually every element in existence, including some that would be better left in the laboratory.'[9] Janine Benyus (author of *Biomimicry: Innovation Inspired by Nature*) states that: 'life builds from around five commonly used polymers'.[10] From these polymers, nature has created a myriad of complexity from Abalone shells that form a material stronger than the toughest ceramic,[11] through to materials such as cotton, valued for its softness and flexibility.

Nature uses structure and form to create stiffness, flexibility and even colour from its limited palette of materials. Structural elements such as bones and tree trunks use complex forms to create light, efficient structures from microstructures that start at the molecular level. A butterfly's wing is coloured not by pigmentation but by a microstructure that refracts light.

Additive manufacture, or 3D printing, allows components to be built up in layers using industrial robot technology created from CAD files or 3D scanners. The technology uses a range of materials including plastics, ceramics and metals to create forms with complex internal geometry. Designers and contractors are now experimenting with 3D printing of building components using traditional construction materials such as concrete, aluminium and glass.

The technology represents both an opportunity and a threat to the circular economy. The main threat is that there is the potential to create a highly complex mix of materials that would be impossible to separate at end-of-life. The idea of printing houses from cement blended with recycled rubble, fibreglass, steel and binders, as demonstrated by a construction firm in China, leaves the legacy of a solid chunk of composite material that will have to be demolished and crushed at some point.

The opportunities, on the other hand, are exciting. Additive manufacture means that components can be shaped and formed through addition, rather than the traditional industrial approach that involved removing material to create the desired design. Additive manufacture avoids the 'yield losses' associated with off-cuts, machining, components rejected due to defects and so on. Research by the University of Cambridge shows that the global yield losses from turning liquid metal into a fabricated product are 26% for steel and 41% for aluminium, so there is the potential for considerable savings.[12] This could be combined with the idea that products can be made with the exact amount of material required to perform the function required, so structures could have material added at points of increased stress and even designed with the complex internal geometries demonstrated in nature.

The potential applications of 3D printing inspired by biomimicry are explored in Michael Pawlyn's book. This includes the idea inspired by the butterfly wings mentioned above: creating a nanostructure from glass that performs a similar function to the low-emissivity coatings for high performance façades. This would mean the glass element would be made from only one material, making it far easier to recycle at end-of-life. Applying this approach of using structure to create different performance characteristics of materials, rather than using more polymers and more ingredients in materials, should contribute towards a more circular economy.

PolyBrick, the Mortarless 3D Printed Bricks

Sabin Design Lab at Cornell University and Jenny Sabin Studio have developed PolyBricks, which are 3D printed bricks that are lightweight, need no mortar and can be designed with just the right amount of material required to meet their function (see Figure 10.06).

The team's inspiration comes from traditional Japanese joinery (as discussed in Chapter 3) and the bricks use a dovetail joint that is elegantly designed to lock the joint into place by sloping the surface so that each piece is held in place by its own weight. There is no need for mortar or adhesive, allowing the assembly to be disassembled and reused. This contrasts with conventional, modern brick and mortar construction where the bricks are bonded together by mortar that is stronger than the bricks, meaning the bricks cannot be reclaimed.

The bricks are printed in clay using a powder-based 3D printer, then fired, dipped in glaze and then fired at a higher temperature to turn them into ceramic. The printing process means that there is no waste material and the shape can be created exactly according to the geometric design.

Figure 10.06: PolyBrick – mortarless 3D printed bricks.

10. SELECTING MATERIALS AND PRODUCTS

The geometry of each brick can be customised to respond to the requirements of a given section of a wall, so a section with additional structural loading can be strengthened, spaces can be created for building services, and openings can be created to allow in daylight. As Jenny Sabin says: '3-D printing allows us to construct and design like nature does, where every part is different, but there is coherence to the overall form.'

The PolyBrick concept uses less material, allows deconstruction and reuse whilst providing an elegant and complex design solution.

ClickBrick, the Mortarless Bricks

A commercially available version of 3D printed interlocking, mortarless bricks may seem a long way off, but the problem of mortarless bricks has already been solved in the Netherlands by Daas Baksteen. Daas ClickBrick® is a system of engineered bricks that are connected to one another using stainless steel clips inserted into grooves at the corners of each unit. The wall is then tied back and supported by the inner leaf using wall ties (see Figure 10.07).

Figure 10.07: Construction of a wall using mortarless bricks.

This provides a system that is completely demountable and the bricks can be reclaimed for reuse without having to be cleaned. Doing away with mortar has several other advantages:

- it increases the speed of construction
- the work can be done in bad weather, and
- it requires less skilled labour.

The finished wall looks clean and simple without the mortar joints, there is no risk of efflorescence and it does not require re-pointing.

Finishes and Fixings

Finishes can inhibit both the maintenance and upgrade of buildings and the deconstruction and reclamation of materials or components. Carpets represent a significant environmental impact as they are frequently replaced long before the end of their service life and a high proportion of the waste is not recycled.[13] Using carpet tiles in commercial buildings is common practice and allows layouts to be changed, worn out tiles to be replaced and access to services under raised floors.

Carpets are a good example of how the principles of designing for adaptability and deconstruction can be applied, by removing the need to glue them down. Carpet manufacturers have developed systems to eliminate the need for adhesives, e.g. there are high-friction lightweight coatings and adhesive-free corner pads to bind the tiles together.[14] These systems allow carpet tiles to be lifted easily for access, replacement or removal at end-of-life. Similarly, there are ceramic floor tiling systems that are laid onto a rubberised grid system. The system avoids any need for grout or adhesive, does not off-gas fumes and it can be installed over an existing floor. It also allows cracked tiles to be easily replaced and they can be reclaimed at end-of-life.[15]

Using materials that have an 'inherent finish' helps to maintain the purity of the material. Timber such as oak, larch or western red cedar can be used untreated for external cladding as it weathers naturally. Aluminium, stainless steel or Corten steel are inherently resistant to weathering.[16] There are timber treatments that extend the life of the wood and allow it to be used externally whilst not using toxic chemicals or contaminating the material, such as the heat treatment used by Superuse Studios (see Chapter 11) and Accoya. As Pawlyn explains in *Biomimicry in Architecture*, both techniques make the wood indigestible to microbes. Thermally modifying wood (e.g. ThermoWood treatment) exposes the timber to intense pressure and heat and Accoya uses a process of acetylation using a benign chemical (acetic acid).[17] For internal finishes, using inherent or natural finishes that do not off-gas air pollutants, such as Volatile Organic Compounds (VOCs), both benefits the internal environment and avoids pollution of finishes during manufacture, application and at end-of-life.

Certification Systems

Selecting materials and products that are compatible with the principles of a circular economy is a complex and involved task. The shortcut is to use certification schemes that analyse and assess products against these criteria and label products according to their relative environmental performance.

There are many certification schemes that consider different aspects of the impacts of materials and products. Some consider the carbon emissions associated with products, while others focus on products with low chemical emissions that may pollute the internal environment.

ISO 14025 sets standards for voluntary environmental labels and identifies three types of labels ranging from third party labels that consider the whole lifecycle through to self-certified declarations of performance.

Examples of labels that align closely with the circular economy principles include the Cradle to Cradle (C2C) Certified™ Product Standard and Natureplus.

Natureplus includes criteria for assessing products against specific criteria for each product group. The criteria are focused on: 'The protection of limited resources by the minimisation of the use of petrochemical substances, sustainable raw material extraction/harvesting, resource-efficient production methods and the longevity of the products. Therefore, building products made from renewable raw materials, raw materials which are unlimited in their availability or from secondary raw materials will be favoured for certification'.[18] Natureplus requires the manufacturer to make a declaration of all the input substances and proof of origin of all the input materials. It includes a list of banned substances that are damaging to health or the environment. It also sets out rules for deconstruction and recycling at end-of-life very similar to those set out in Chapter 9.

Cradle to Cradle (C2C) Certified™ product

The Cradle to Cradle (C2C) Certified™ product was originally developed by McDonough and Braungart and aims to implement the cradle to cradle philosophy (see Chapter 1). The scheme assesses products under the following categories:[19]

- Material Health: knowing the chemical ingredients of every material in a product, and optimising towards safer materials.
- Material Reutilisation: designing products made with materials that come from and can safely return to nature or industry.
- Renewable Energy and Carbon Management: envisioning a future in which all manufacturing is powered by 100% clean renewable energy.

- Water Stewardship: managing clean water as a precious resource and an essential human right.
- Social Fairness: designing operations to honour all people and natural systems affected by the creation, use, disposal or reuse of a product.

Under the standard, products are awarded a score of: Basic, Bronze, Silver, Gold and Platinum, showing the level of achievement against each of the above categories. This provides manufacturers with the opportunity to achieve an entry-level certification and a pathway to how they can achieve the highest rating.

Cradle to Cradle® Certification includes a list of chemicals that are banned due to their tendency to accumulate in the biosphere and lead to irreversible negative human health effects. These are split into technical and biological nutrients. The separate lists allow for the use of some substances, such as lead or cadmium, to be used in materials where it is unlikely that there will be exposure to humans or the environment. For example, lead is used in cast aluminium, but it does not migrate out of the material. However, lead should not be used in biological nutrients, to ensure that it is not released into the biosphere.[20]

Figure 10.08: NASA Sustainability Base.

10. SELECTING MATERIALS AND PRODUCTS

NASA Sustainability Base

The NASA Sustainability Base in California (completed in 2012) was designed by William McDonough + Partners with AECOM as the architect of record and engineer of record (see Figure 10.08).

A rigorous selection process was applied to the choices of materials and products that were used on the building to implement the cradle to cradle vision for the project. Cradle to Cradle Certified™ products were used when available and when they were deemed to be cost effective. The other products were evaluated by McDonough Braungart Design Chemistry (MBDC) using the Cradle to Cradle® criteria. Figure 10.09 shows the Cradle to Cradle® products that were selected, including elements of the cladding, the hardtops including elements of the cladding and the hardtops.

Figure 10.09: NASA Sustainability Base, Cradle to Cradle Certified™ products.

- Centria Dimension Series® panels (certified silver)
- Alcoa, Inc. Kawneer 1600 SunShade® louvers (certified silver)
- PPG Industries Solarban 70XL™ architectural glass (certified silver)
- Alcoa, Inc. Kawneer 1600 Wall System® (certified silver)
- Alcoa, Inc. Kawneer InLighten® Light Shelf (certified silver)
- Mechosystems, Inc. Mecho®/5 with EcoVeil (certified silver)
- Icestone® Durable Surface (certified gold)

Cradle to Cradle Certified™ is a certification mark licensed to Cradle to Cradle Products Innovation Institute

97

Figure 10.10: NASA Sustainability Base showing steel 'exoskeleton' structure.

The main building elements are formed from steel, glass and aluminium and are selected with a high recycled content and their ability to be reclaimed or recycled. The lobby areas reuse oak flooring from a transonic wind tunnel on the NASA Ames Campus.[21] The steel structure is designed for deconstruction (see Figure 10.10) and the exterior cladding consists of prefabricated modular units.

The Cradle to Cradle (C2C) Certified™ product is gaining traction in the construction industry with recognition in two environmental assessment methods: RICS Ska Rating and LEED v4. As it gains in momentum, the number of construction products that are certified is increasing, which provides more opportunities for designers to specify compliant products.

Whatever eco-labelling system is used to help with product selection, the main consideration should be to check that it is third-party verified and that it considers the whole life of the material or product, including the manufacturing and in-use impacts and the ability to disassemble, reuse or recycle at end-of-life.

Conclusion

Materials and component selection forms the bedrock of a circular economy. Materials should be selected based on the following:

- Decide whether the element will have a long or short lifespan, in reality, not according to a table, and select the appropriate material that matches that lifespan.
- Determine whether the materials are technical or biological. This decision depends on the lifespan and the likely fate of the element as well as its function.
- Retain the purity of the materials by avoiding mixing biological and technical substances and ensure that biological materials are not contaminated by toxins that inhibit them being returned to the biosphere.
- Select reused and reclaimed components and materials.
- Ensure that elements can be disassembled at end-of-life with the ability to upgrade or remanufacture or reclaim the materials.

These decisions are complicated and require a focus on procurement and a lot of information about the substances in materials and components. Labelling and rating schemes can shortcut many of the complexities in these choices by vetting and rating components for designers. This level of scrutiny on materials makes it essential to build partnerships and collaborative relationships with the supply chain. This is explored further in the case study in Chapter 13 (Park 20|20).

By implementing these ideas, buildings can be turned into materials banks that allow future generations to salvage valuable materials. Equally, existing waste streams can be turned into valuable materials that can be salvaged and used in creating new designs. Turning waste into a resource is explained in the next chapter.

11. Turning waste into a resource

The goods of today are the resources of tomorrow, at yesterday's resource prices.

Walter R. Stahel

In a circular economy, waste becomes a resource. Current demolition practices are driven by the pressure to tear down old buildings quickly and safely, and the lack of demand for reused components means that the majority of materials are downcycled. By designing buildings for disassembly and deconstruction, the residual value of an obsolete building can become positive and there is more incentive to reclaim components and materials.

Redefining waste as a resource requires systemic changes to the industry. To start with, building design has to enable materials to be reclaimed by designing for deconstruction/disassembly and materials in the building have to be catalogued so that the residual value can be calculated and new markets for components can be established. When the building is due to be demolished, more time, space and labour have to be allowed for deconstruction. Lastly, there needs to be an active market for salvaged components and materials that covers the costs of disassembly, storage and resale. Creating a market for salvaged stock means that reused components have to be as readily available, attractive and as fully certified as new products. And there has to be a change of mindset in designers to tailor their designs to incorporate reclaimed materials and components in their new designs.

This may seem impossible, but other industries are taking steps down this path and the case studies in this book show how some building projects are starting to think differently by using reclaimed products in building design and fit-outs, and by designing for deconstruction.

Raw Materials Passports

Thomas Rau is an architect based in Amsterdam. As well as being an advocate of service models over ownership (see Chapter 12), Rau proposes that all the components of a building should have a 'raw materials passport' to give waste an identity. As he says: 'Waste is a raw material without identity.' The idea behind this term is that a 'passport gives you identification – you can't leave the country without it'.[1] If materials do not have a description, then they are likely to be downcycled and much of the value will be lost.

One operator in the shipping industry has started developing materials passports. Maersk Line shipping is dependent on steel to build its ships and has recognised that improved recycling of its decommissioned ships would protect itself against volatile steel prices. When ships are disassembled, it has not been possible to differentiate between different grade materials, meaning that the recycled material ends up being of a lower grade.

Maersk has set itself a task to separate out the high-grade and low-grade steel, as well as the copper from its ships, to ensure that it maintains the value of the materials at the highest possible grades. To do this, it is developing a materials passport to describe all the materials used to build its Triple E ships with information on how to disassemble and recycle them. It is creating a database to capture information on all the materials used and their locations, in a 3D model.[2] BIM offers the same opportunity in buildings, by allowing constructors to create detailed information about the ingredients of a building and communicating that to operators and those responsible for refurbishing or dismantling the building.

Materials Banks

Once the raw materials are all identified and catalogued, Rau proposes that buildings can become materials banks, where materials can be deposited and used in the building, but can be later withdrawn and used elsewhere whilst still retaining their value for use in the future.

Brummen Town Hall, Brummen

Thomas Rau designed Brummen Town Hall to be a 'raw materials depot'. The brief called for a building with a 20-year life so RAU Architects, in partnership with BAM, proposed a radical solution that included using the existing building and wrapping a new structure around it with the idea that the extension could be disassembled and reused at the end of its 20-year life.

Completed in 2013, the building uses prefabricated timber elements for the structure, façade and floors to facilitate reclamation and reuse. The use of concrete was reduced as far as possible, with the retaining walls around the perimeter using recycled concrete held in gabion cages to allow disassembly. All of the suppliers have contractual arrangements to take back their products when the building is disassembled.

This design approach prompted discussions with the manufacturers and suppliers to find out how the building elements could be designed for recovery and reuse. Notably, the timber structure uses timber beams that were made slightly larger than required so that when the building is dismantled it will be easier for the supplier to put them back on the shelf and resell them.

Rau also implemented the Turntoo principles that he has developed (see Chapter 12) to incorporate service contracts for many of the fixtures and fittings in the building, including the internal partitions (Interwand), lighting (Philips), chairs (Steelcase) and flooring (Desso). The elegant main reception desk in the lobby is made from cardboard, allowing it to be returned to the biosphere at the end of the building's life.

Creating a Marketplace for Salvaged Materials

A project that was developed during the UK-Green Building Council (GBC) Future Leaders programme[3] builds on these ideas by proposing an online hub to create a marketplace for building materials (see Fig 11.01). The hub would document the materials used in the building and make the information accessible to interested parties. The materials would be documented in a materials passport which would contain a quantified list of construction materials and modular elements (e.g. windows) along with their recovery rating.

The hub would allow users to find out when materials are available and place bids on them. It could help to assign a value to materials held within existing buildings, with interested parties being able to purchase an interest in the material in advance of it being salvaged from the building.

The UK-GBC Innovation team argues that building owners often know that the building is to be demolished well in advance of its actual demolition, but the actual demolition time is usually very constrained allowing little or no time to find a market for salvaged materials. Selling the materials in advance could incentivise the building owners to allow more time to deconstruct the building and ensure that the components are salvaged intact. In the long term, the hub could also help to encourage the design of buildings that are simple to deconstruct, to allow valuable components and materials to be reclaimed at end-of-life. Creating a guaranteed market for reclaimed materials

BUILDING REVOLUTIONS

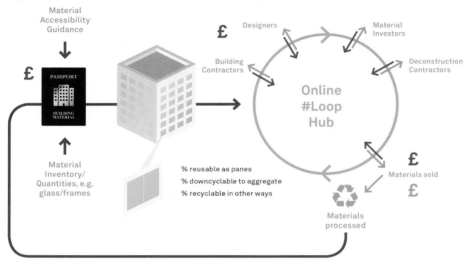

Figure 11.01: An online marketplace for materials, proposed by UK-GBC.

prior to demolition would also provide an economic incentive for demolition contractors to salvage components and materials.

Creating a Network

There is an opportunity to salvage materials and components from other industries by creating networks and exchange forums that allow products to be brought to the attention of the relevant people. For example, through some research, the project team for the London Olympic Park were able to source surplus gas pipelines from the Statoil's Langeled project (bringing gas to the UK from Norway) to form part of the roof structure of the main stadium.

As these networks and exchange forums are rarely available or well-populated, some entrepreneurial organisations are creating their own platforms, using technology to locate and share resources.

Superuse Studios, Rotterdam

Superuse Studios, an architectural practice based in Rotterdam, has turned reclaiming materials into both an art and a science. It has put together a multidisciplinary team, including chemists, environmental scientists and analysts, and its aim is to 'help designers turn cities into a living web of connected material processes and flows'.

Figure 11.02: Harvest map.

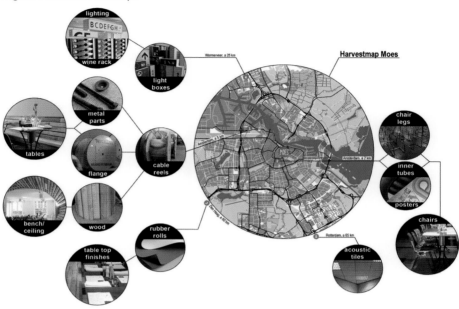

The science is in the application of technology to source and treat the materials. Superuse Studios has created a whole suite of web platforms to enable connections between material demands and supply. Reclaimed materials are sourced with the help of a 'harvest map' web platform (http://www.harvestmap.org) that provides a library of materials and their locations (see Figure 11.02). It is populated by a team of 'scouts' who find and procure the materials using resources such as Google Earth to locate potential stocks of materials. These web tools can be used to help designers find waste materials close to the site of the project and they provide guidance on how to utilise the resources.

The materials can then be processed to ensure that they can be reused. As Jan Jongert, founding partner and Head of Research at Superuse Studios, says: 'The idea is to create new value from what is already available and to create new flows of materials.'[4] Jongert warns against creating closed materials cycles, as manufacturers who keep the materials within their own control will stop other suppliers having access to these resources. Rather than a closed-loop approach, he is more interested in the idea that waste from one industry becomes a resource for another.

Superuse Studios is working with Plato Wood, a company that has found a way to reuse and upgrade the properties of timber. Superuse Studios is using the process to

Figure 11.03: Cable reels.

upgrade the quality of the timber that they reclaim from the ubiquitous cable reels that are a common sight around the industrial fringes of Rotterdam (see Figure 11.03). The softwood from the cable reels is heat-treated to upgrade the low-grade timber into a material with a lifespan of up to five times longer with good dimensional stability, improved sound insulation properties and improved strength.

The timber is heated, dried, cured and conditioned using steam from the adjacent co-generation plant, a good example of industrial symbiosis in action. No chemicals are used in the process and the treated material can be used without adding any additional coatings.

The art, after the science, is to find new and creative architectural uses for the reclaimed materials. Superuse Studios designed Villa Welpeloo, in Enschede, a house and a private art gallery, to be made predominantly from reclaimed material (see Figure 11.04). Completed in 2010, the building is clad in the heat-treated timber reclaimed from the cable reels.

The main structure uses steel profiles that were reclaimed from a textile production machine (see Figure 11.05).

The idea that material that would otherwise have been scrapped or burned can be reclaimed and 'upcycled' into a higher value product is exciting. This case study shows how online resources can be used to create networks between different industries and it offers opportunities for entrepreneurs to set up businesses and mediate between the different industries to find new markets and uses for waste materials.

11. TURNING WASTE INTO A RESOURCE

Figure 11.04: Villa Welpeloo.

Figure 11.05: Reclaimed steel beams.

Conclusion

Turning waste into a resource means stopping it becoming waste in the first place. This means that materials have to be clearly identified, catalogued and separable to allow a value to be attached to them. Designers have to think about using reclaimed materials and how they can be salvaged at end-of-life, contractors have to be able to source materials and components that are comparable to new products, and demolition contractors have to become brokers who salvage and sell resources. This can only work if new, vibrant networks are set up to make links between different industries and create a flow of materials that keeps resources in circulation. The circular economy needs a hub around which to turn.

12. Circular business models

In order to change an existing paradigm you do not struggle to try and change the problematic model. You create a new model and make the old one obsolete.

Buckminster Fuller

The linear economy is based on manufacturing products, selling them to a consumer, who then consumes them and is responsible for disposing of them at end-of-life. The circular economy needs new business models to retain the value of products and to help to close the loop.

The business models range from full performance-based models through to take-back and remanufacture. In the performance-based models, the manufacturers retain ownership of the products whilst providing a service to the customer. The take-back model proposes selling products that are designed to be returned and remanufactured with incentives such as deposits, or guaranteed buy-back, to encourage customers to return the used products when no longer required.

Performance not Products

The industrial economy is transforming from a production-based model into a more intelligent performance-based model. Yet despite the proven benefits that selling performance provides, too many managers and policy makers still focus on designing, manufacturing, and selling goods using costly economic models and production methods.

Walter R. Stahel

Walter R. Stahel has long promoted the concept of selling performance rather than products, so the consumer becomes a user and ownership is replaced by stewardship. There is an increasing shift towards selling performance over purchasing products, with people leasing printers, cars, office space and even clothes (e.g. Mud Jeans). Selling performance means that the manufacturer retains ownership of the materials, therefore securing their supply of components and materials in the future. This helps to maintain the value of the components, as manufacturers are more likely to:

- design their durable products to have a long service life with the ability to repair and upgrade, or even remanufacture them
- design their consumable products to be simple to disassemble and recycle, or to return to the biosphere
- ensure that toxic materials are easy to separate and reclaim for reuse, or are designed out of the products.

These arrangements can be good for businesses, as they create long-term relationships between manufacturers and customers, rather than one-off interactions.

RAU Architects and Turntoo

Thomas Rau's design philosophy rests on the premise that the planet is a closed system and that there are a finite amount of resources available with no new places to put the waste that is generated. His designs aim to make best use of materials and resources to respect the balance between humans and the physical environment. Rau proposes that humans should behave as guests on the planet, rather than owners of it. One way to realise this in building and product design is to shift away from ownership and towards stewardship. Owning products brings responsibility for their upkeep and ultimate disposal, which is problematic when many products are full of materials that are unidentified and difficult to reclaim as an equally valuable element.

As part of the drive towards optimising the use of materials, Rau has implemented the use of service models and as well as leasing services in some of his buildings (see Brummen Town Hall case study in Chapter 11), and he has set up a company called Turntoo that facilitates service-based contracts between manufacturers and users. The idea of Turntoo is that the manufacturer retains ownership of the products and the consumers only pay for the performance, rather than the raw materials that go into the product. This creates a closed loop with the materials and components of the product remaining in a 'raw materials cycle' whilst the service offered by the product is agreed with the customer.

Based on this concept, Rau approached Philips lighting and proposed that they offered him a lighting service instead of purchasing lamps and control gear. The idea developed into Philips' 'Pay-per-lux' model, which was pioneered in RAU Architects' own offices in

Amsterdam and subsequently in Brummen Town Hall. The Pay-per-lux model is a performance-based arrangement where Philips installs and maintains a lighting level and monitors the lighting performance and energy use online with annual reporting, health checks and preventative maintenance. Furthermore, Philips pays the energy bills for the lighting, which gives them the incentive to provide the most efficient lighting system and to ensure that it is operating as well as possible.

As Thomas Rau says: 'I told Philips, "Listen, I need so many hours of light in my premises every year. You figure out how to do it. If you think you need a lamp, or electricity, or whatever – that's fine. But I want nothing to do with it. I'm not interested in the product, just the performance. I want to buy light, and nothing else."'[1]

The model was re-conceived by Philips in the UK specifically for the National Union of Students.

Macadam House, London – Pay-per-lux

When the National Union of Students (NUS) set about refurbishing an old 1960s building into its new headquarters in London, it wanted to embed some of the principles of the circular economy into the project, as well as incorporating best practice sustainability features (such as photovoltaics, green walls, etc.).

The NUS works with student unions throughout the UK to educate students about sustainability and it runs several high-impact campaigns aimed at changing behaviour. The theory is that when people are going through a significant change in their lives, such as leaving home to become students, it is a good time to instil new behaviours into them. The NUS wanted to embody sustainability principles into its refurbished building (completed in 2013) to show what was possible.

Jamie Agombar, the Head of Sustainability for NUS, asked the suppliers and manufacturers on the fit-out contract whether they could offer a service instead of selling their products to the NUS. The only manufacturer that came back with an offer was Philips lighting. Philips proposed a model where it provided a lighting service, rather than selling the light fittings – its Pay-per-lux model.

To deliver the service, Philips provided and installed around £120,000 of LED lighting with comprehensive controls throughout, instead of the less efficient T5 (fluorescent) lamps that NUS would have had to install with its original £40,000 budget (see Figure 12.01).

The NUS pays a quarterly rental payment to Philips for the service, with pre-agreed rates that incentivise efficiency. The energy use of the lighting was estimated at the start of the contract and agreed between the NUS and Philips. If the energy use exceeds this figure, then the NUS pays less rent for that quarter.

Figure 12.01: Macadam House, NUS Offices, London, showing Pay-per-lux lighting.

'As a registered charity we didn't want to own services like the lighting,' Agombar said. 'Our priority was to ensure the lighting performed as required in terms of light levels and energy consumption.'

To ensure efficient performance, Philips has trained the staff in the building and has been fine-tuning the controls systems. The system now works far better than typical lighting systems in buildings, where sensors are often not recalibrated or settings are overridden.

The fact that Philips is providing a service and retains ownership of the equipment makes three contributions towards the circular economy philosophy:

1. Philips is responsible for any upgrades and replacements to ensure that it retains its performance and rental income. This should incentivise Philips to design its systems to be easy and cost-effective to upgrade.
2. It is responsible for the fate of its equipment at end-of-life, meaning that it should be designed for disassembly, remanufacture or recycling.

12. CIRCULAR BUSINESS MODELS

3. Philips is effectively using the building as a 'materials bank' for the future, helping it to secure the source of its components and raw materials in the future. As an added benefit, the energy target incentivises Philips to maintain the lighting to ensure it is performing to its optimum efficiency.

The NUS is now letting out two floors of the building and, as a landlord, is providing the Pay-per-lux as a service to its tenants, in the same way that it provides heating, cooling and ventilation. This opens up the possibility that landlords could provide a lighting service to their tenants or act as a broker for a lighting manufacturer.

Philips has subsequently developed a lamp prototype that can be disassembled and reassembled without any need for tools or adhesives. This could allow the electronic board of the light source to be upgraded or the parts of lamp to be reused or recycled separately.[2] This shows that creating new ownership models can incentivise manufacturers to redesign their products for a more circular economy.

Incentivising Return

Selling services instead of products is one way to ensure that building components and contents are returned to the manufacturer for reuse or remanufacture. Another way to close the loop is to incentivise customers to return products when they are no longer required by offering to buy them back or to provide a discount on the next service.

Some organisations have recognised the advantages of ensuring that their products are returned and have set up the infrastructure to collect products, the incentives to ensure a high return rate and remanufacturing capabilities to squeeze the maximum value out of the used products.

Caterpillar manufactures machinery and engines, including construction and mining equipment. It uses its vendor and distribution system to collect used components and operates a deposit and discount system to incentivise their return.[3] Customers pay a deposit when they purchase new equipment, which is paid back when they return the product. Caterpillar operates a remanufacturing division that disassembles the products, then cleans and inspects the components to determine what can be salvaged. The components and the 'core' of the machinery or engine can then be remanufactured to a very high performance standard, which allows Caterpillar to offer the same warranty for their CAT® Reman engines as for new products.[4] This process allows Caterpillar to:

- get valuable feedback on the design of their products
- reduce their materials costs
- provide a reliable source of remanufactured components
- provide a parallel business stream of remanufactured products and
- help to create long-term relationships with their customers.

113

This business model is starting to be applied to building components and contents, such as furniture.

Rype Office, London

Furniture reuse is a well-established practice in the UK, with networks and organisations offering reclaimed furniture. However, WRAP estimates that only 14% of office desks and chairs are reused every year in the UK with the remainder going to landfill, energy recovery or recycling.[5] This equates to around 75,000 tonnes of office furniture going to landfill annually.[6]

Figure 12.02: Remanufacturing process.

Rype Office draws on circular economy principles to create high quality furniture with less environmental impact, whilst creating local jobs through remanufacturing. Its business model addresses every stage in the life of furniture with the aim of moving beyond the linear model of consumption and disposal. Rype Office helps companies to refresh and resize their existing furniture, remake furniture sourced from elsewhere at half the cost of new and choose new furniture that is easy to remanufacture, giving it a much longer potential lifetime.

It also offers guaranteed buy-back on its furniture and even a leasing service, giving customers the flexibility to return unwanted furniture or to lease more as their organisations change.

According to Dr Greg Lavery, director of Rype Office, the most common reasons for not using remanufactured furniture are concerns about quality and about the volumes that can be supplied.

'When you put top grade remanufactured furniture alongside new you cannot tell the difference, not even with a magnifying glass – that is how good it is. Modern resurfacing technologies and remanufacturing processes are very high quality. And of course you can choose the colours and finishes that you want because we are completely remaking it,'[7] says Greg in response to the first concern.

And in response to the second concern about whether the volume of stock is available, he notes that there are hundreds of thousands of furniture items from office clearances now sitting in warehouses since the global financial crisis, just waiting to be remanufactured.

Rype Office's remanufacturing model creates:

- remade furniture at half the cost of new
- reduces the environmental footprint of each piece by an estimated 70%
- creates skilled jobs in the UK, and
- reduces the amount of new furniture and components that have to be imported into the UK.

Conclusion

These service models could be extended to encompass the whole fit-out of a building, allowing occupants to lease everything they need from the manufacturer. This could give occupants access to the latest technology or trends and allow them to transfer the risks of trialling these new ideas to the manufacturer, along with the total cost of ownership.

At first glance, leasing models appear more expensive than simply purchasing products outright. However, this can change when the total cost of ownership is considered. Calculating the total cost of ownership involves gathering together a whole range of costs that are not normally associated with the purchase of the equipment. This includes the costs associated with maintenance, premature failure, functional obsolescence, replacement and the storage of surplus equipment or replacement parts. Depending on the leasing models, it could also include the energy costs of mechanical and electrical equipment and the disposal costs of products, including hazardous materials that would require specialist handling.

These business models help to ensure that products and materials are kept in circulation for longer and also offer multiple benefits to manufacturers and customers, if they are set up to the benefit of both parties and the right infrastructure is put in place. Manufacturers can develop longer-term relationships with their customers, whilst reducing their cost of materials and their exposure to volatile materials prices. They can create a second revenue stream and customer base by selling remanufactured products and obtain feedback on their products, allowing them to develop higher-quality, longer-lasting goods. Remanufacturing creates local employment opportunities and reduces the need to import goods. The customers benefit from lower-cost products with the potential to enter into service agreements that mean they are not responsible for the maintenance, upgrade and disposal of products.

13. Virtuous circles

If humans were to devise products, tools, furniture, homes, factories, and cities more intelligently from the start, they wouldn't even need to think in terms of waste, or contamination, or scarcity. Good design would allow for abundance, endless reuse, and pleasure.

William McDonough and Michael Braungart

Innovative design and creative thinking needs the right environment to flourish. Engaging with the supply chain at the back end of the design process, once all the main decisions have been made, sounds wrong. To then beat the suppliers down to the lowest cost through competitive tendering is not going to create a nurturing environment for new ideas.

Engaging and partnering with the supply chain is standard practice in other industries and leads to strong, trusting relationships and creates the space to innovate.

Moving away from lowest-cost tendering sounds expensive, but Coert Zachariasse of Delta Development, based in Amsterdam, has made it work (see case study below) by implementing an open-book accounting system and by asking the suppliers to provide the best product they can for the allocated budget. In return, they become the guaranteed suppliers for a series of building projects. This creates a fertile environment for design innovation, avoids construction delays and creates a better building that commands more rent.

Zachariasse has also demonstrated at Park 20|20 (see in more detail, below) that it is possible to design buildings using circular economy principles. The buildings are designed to be more adaptable to reduce the risks of obsolescence, the careful selection of materials and components has led to a healthier internal environment, and the buildings should have a residual value by designing for disassembly.

Park 20|20

Park 20|20 is a new business park one train stop from Amsterdam's Schiphol Airport. It is the first business park in the Netherlands to be inspired by Cradle to Cradle® thinking and aims to implement many of the principles of the circular economy. Park 20|20 is the brainchild of Delta Development's CEO, Coert Zachariasse.

Coert Zachariasse's journey to creating Park 20|20 started in 2004 when he was demolishing buildings on the redundant Fokker airfield factory site in Amsterdam. He realised that there were valuable materials on the site that would be wasted through conventional demolition practices, so he initiated a process whereby he retained ownership of on-site materials which were subsequently deconstructed and reused. Four years later he was inspired by a presentation given by William McDonough on the Cradle to Cradle® philosophy (see Chapter 1) and decided that he wanted to apply the principles to a new business park. He obtained the land, the support and the funding that he needed based on this vision. Figure 13.01 shows the masterplan for the park.

Zachariasse wanted to all the products obtained for the building to be Cradle to Cradle® certified, but in meetings with McDonough it quickly became apparent that there were not sufficient products that had been certified at the time to build a whole building. Rather than water down his goal, McDonough and Zachariasse developed a pioneering approach to drive the supply chain to respond. The team scanned 350 different products for their potential to be Cradle to Cradle® products and rated them red, amber and green. They talked to the amber companies and asked them to phase out the materials or chemicals in their products to allow them to be certified. This resulted in 41 suppliers who either had C2C certification or who had been vetted to provide an acceptable standard of product.

Zachariasse then changed the whole tendering process to provide a direct relationship with the manufacturers and suppliers. Firstly, he made the contractor a partner in the project and created an open-book accounting system. He asked them to break down their costs into: direct costs (materials and labour), site and construction costs, and the profit margin. Zachariasse then offered to double their profit margin in return for total transparency on the direct costs. Between them, they prepared a budget for each of the elements of the building (façades, structure, upper floors, etc.) and then asked the suppliers to provide them with the highest quality product that they could for the cost. This approach encouraged the suppliers and manufacturers to propose innovative solutions in the early design of the project whilst giving Zachariasse control over the supply chain. The manufacturers knew they were the suppliers for the project and so could prepare and manage their supply chains accordingly. This meant that there were

13. VIRTUOUS CIRCLES

Figure 13.01: Park 20|20 masterplan.

no delays or issues with materials and products not being delivered on time, saving time and money on site. BIM was used to create further savings in both construction costs and materials procurement.

Turning the procurement process on its head like this is in sharp contrast to the traditional procurement route where suppliers are asked to tender for work after all the design decisions have been made. It means that developers, designers and contractors can work collaboratively with the manufacturers to create new solutions that use fewer materials, waste less and provide quality solutions.

Overall, the shortened construction time and reduced supply chain costs saved 18% on the project. The same team has been retained for the other buildings on the business park, which has given them security, whilst keeping the consistency and allowing them to learn and improve the design. Many of the same building elements, supply chain and detailing are used on the new buildings, with refinements in the design as lessons are learned. This helps to address one of the fundamental problems with building design and construction: each building is an untested prototype. By integrating the supply chain and repeating the design, the lessons from the first building are fed into the next and the design effort is not wasted. This makes it sound like all the buildings will be the same, but of course this is not the case as the building system allows for different

Figure 13.02: ANWB Reizen building with part of Bosch Siemens building in the foreground.

façades to be used, different configurations of spaces and different geometries, as shown in Figure 13.02.

Delta Development is moving beyond the traditional developer role. It works closely with the tenants to develop spaces that match what they actually need rather than what it thinks they want. Zachariasse says that 'real estate starts with metrics such as floor area, when it should be starting with values and a vision, developing goals, objectives and then settling on metrics'.[1] When Bluewater said it wanted 11,000m^2 of office space on the 20|20 Park, Delta discussed this and they worked out that their current working models meant they needed around 6,000m^2. They settled on 8,000m^2. Zachariasse notes that the role of offices is changing; they are now for meeting colleagues and places to find inspiration rather than just places for desk-based work. This means that organisations often need less or different types of space (see Figure 13.03).

Delta also fits out the building to provide a 'turnkey' solution as this gives it control over the choice of materials for the interiors. As part of the fit-out, Delta has entered into leasing arrangements for some of the elements. There are leasing contracts with LED Lease and BB Lights for the lighting, office furniture is leased from Ahrend and the carpet tiles are leased from Desso. Delta is also working with HM Ergonomics, which is researching a full interior leasing concept, including partition walls.

13. VIRTUOUS CIRCLES

Figure 13.03: Bluewater building, interior.

The results are impressive. The buildings at Park 20|20 are all fully let on completion and Zachariasse's figures show that they are commanding 'a 70–80% higher rental than the buildings on the other side of the dyke'.[2] The buildings:

- are designed with flair, despite (or perhaps because of) the limited palette
- have abundant daylight
- have an abundance of internal plantings, and
- use non-toxic materials.

Large, central atria provide plenty of space for circulation, breakout areas from meetings and events, and a sense of space and light. Internal green walls and trees provide contact with nature, enhance air quality and bind fine dust (see Figure 13.04).

One of the main attractions is the quality of the internal spaces and the potential gains in the wellbeing and productivity of the buildings. The Bosch Siemens employees were surveyed by Delta both before and after they moved into the new office; the results of the second survey showed a 5–6% increase in productivity.

The buildings have been consciously designed to allow adaptation to extend the life of the building and for disassembly at end-of-life. Notably, the designers have attempted to solve one of the intractable problems when designing for adaptation and

Figure 13.04: Bosch Siemens building atrium with green wall.

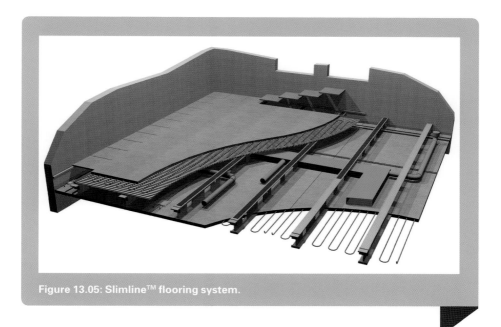

Figure 13.05: Slimline™ flooring system.

disassembly: how to construct the upper floors. Park 20|20 pioneers a flooring system called Slimline™ which consists of a precast concrete ceiling integrated with steel beams and a sub-floor that allows access to the service void created by the steel beams (see Figure 13.05).

The accessible horizontal void allows services to be upgraded to the latest technology without having them poured in concrete. This design feature should also allow for the building to be sufficiently adaptable to accommodate different tenants. When the market changes and offices need to be turned into a different type of building, say a hotel or apartments, the system is designed to allow this adaptation by providing:

- the potential to add (or remove) services within the floor void
- acoustic separation using the double-layer construction
- modular design that allows staircases to be reconfigured by removing whole floor sections without having to break up screed
- a precast finished soffit that has a smooth finish for aesthetic value.

The system, as the name suggests, provides a slimmer floor sandwich than conventional construction and uses considerably fewer materials, providing a weight saving for the whole building. At end-of-life, the steel beams are relatively easy to separate for recycling and the concrete can be, inevitably, downcycled for aggregate.

The C2C certification includes criteria to assess whether the products can be demounted or disassembled (see Chapter 9). This has been used in Park 20|20 to provide buildings that can be redesigned with a new appearance as well as a new function. The Bosch Siemens building includes a demountable, C2C certified ceramic tile façade from Mosa that can be removed and replaced, if a new look is required.

The buildings are designed around an 8 x 8 m grid, which works equally well for commercial or residential buildings in the Netherlands. The structural steel frame is modular and uses standard-sized beams to increase the chances that they may be reused at end-of-life.

Delta asked a demolition contractor to provide a quote for reclaiming the materials on the conventional building design of the Bluewater building on the 20|20 Park. Zachariasse's figures show that there is a €1 million-worth of steel in the building. Using conventional construction techniques the contractor quoted €45,000-worth of scrap metal and €48,000 to extract the metal from the building leaving a negative residual value of €-3,000 to reclaim the steel for scrap. This is an example of how buildings typically have a negative residual value at end-of-life. The use of a structural screed was cited as the main reason for the high cost of extraction and the low value of the salvaged steel. Once the designers had redesigned the building to make it easier to deconstruct, the value of the steel went up by €195,000 with an additional construction cost of €70,000 and additional deconstruction costs of €45,000. This left a net residual value of €85,000, giving the building a potential positive residual value at end-of-life.

Conclusion

Park 20|20 demonstrates that buildings really can be designed for a more circular economy and that the principles discussed in this book can be implemented. The case study shows that designing within a circular economy makes financial sense but, to really work, the construction industry needs to be turned upside down and given a shake.

Designing for adaptability or deconstruction is hard to justify and is unlikely to happen unless it is part of a wider story that starts with reducing construction time on site, continues with the ability to retain value by adapting buildings to changing markets and concludes with the attractive idea of providing residual value rather than demolition costs.

Figure 13.06 summarises the potential changes in revenue and costs associated with designing buildings with a long-term view, based on the ideas proposed by Coert Zachariasse and inspired by a graph drawn by researchers at Loughborough University in relation to adaptable buildings.[3] It compares the conventional building model with a

Figure 13.06: Illustration of the benefits of circular economy buildings, inspired by a graph drawn by researchers at Loughborough University and by the ideas of Coert Zachariasse

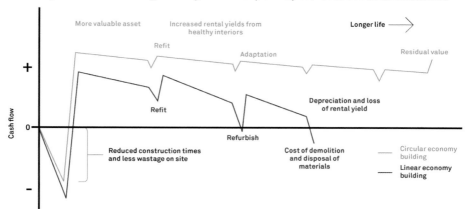

more adaptable one that is designed for disassembly and uses materials that provide a healthier internal environment. The collaborative approach used by Delta Development can reduce construction time, and therefore cost, while the costs of refitting and adapting the spaces, or even the building, should be lower and require less time. There is the potential to have lower rates of depreciation and perhaps to sustain higher rental yields. Finally, designing for disassembly can provide a positive residual value.

Making small, incremental changes towards a more circular economy could drive up capital costs or design fees, which can be hard to defend against the onslaught of lowest-cost tendering and value engineering. Embracing the whole circular economy narrative, on the other hand, means collaboration with the entire supply chain and the ability to design for the whole life of the building, including 'beyond the grave'.

Buildings are a legacy for future generations. There is an opportunity to leave a valuable inheritance of buildings that can be adapted to new uses or deconstructed and turned into new buildings, in whatever form the future demands.

14. Coming full circle

You don't have to do any of this, survival is not mandatory.

Walter R. Stahel

This book demonstrates that circular economy thinking can be applied to buildings and that there are opportunities for those that embrace the ideas to create new business models and designs that enable a transition to a more regenerative built environment.

The cases studies in this book show that there are some visionary and enterprising clients, designers, constructors and manufacturers who are capitalising on the opportunities presented by a circular economy. They have managed to overcome the barriers to change and the inertia in the system to differentiate themselves in the market.

These new ideas respond to the pressing need to slow and reverse the growing demand for raw materials and the related environmental and social impacts. By implementing these ideas, businesses can seize more control over their raw materials and protect themselves from supply disruptions and volatile prices while reducing their exposure to environmental legislation and the costs of waste disposal. Businesses can respond to the shift in the market by providing performance rather than products, allowing them to forge new longer-term relationships with their customers.

Cycling, not Recycling

Embracing the idea of a circular economy is so much more than just reducing waste arising on site, or recycling more waste. It is about fundamentally rethinking the way that buildings are designed, procured and used. It is about avoiding design that takes buildings down blind alleys, where there is no way back except depreciation, demolition and downcycling. It is about creating a built environment that retains the value of buildings, components and materials in a virtuous circle, while making buildings that are better for the environment and for people.

Thinking Beyond the Grave

This book outlines a set of design principles that interpret the circular economy model for the built environment. The trick is to think beyond the immediate demands for the building and to consider the potential next lives of the building and its components. The design principles aim to capture circular economy thinking and to apply it to building design, as follows:

- Design-out waste by refitting and refurbishing buildings in preference to demolition and building new, using waste as a resource to create new designs and using lean design principles to create buildings that use fewer resources and components and are less complex.
- Create structures that are built to last, that have layers that can be peeled off and replaced, that can be adapted to new uses and that can be disassembled and even reconfigured for different uses on new sites.
- Select building components that are carefully designed to flow in either a technical or a biological cycle, depending on their lifespan, use and what is available. The components in the technical cycle are designed to retain value, by being designed for reuse, remanufacture or disassembly. The biological materials remain uncontaminated and can be cascaded through different uses before being returned to the biosphere at end-of-life. Building components and materials in the technical cycle that are hard to salvage at end-of-life should be designed-out or perhaps replaced with biological materials.

Leaps and Bounds

New technologies and business models can leapfrog over current thinking and leave it standing, allowing industries to escape the bounds of convention.

3D printing, or additive manufacture, provides the potential to make beautiful components that do not need complex polymers or irreversibly bonded materials. New designs could be inspired by the way that nature uses structure and form to create stiffness, flexibility and even colour from its limited palette of materials. Additive manufacture reduces waste during production and, if applied correctly, could help to create components made from only a few polymers that are simpler to recycle at end-of-life.

Information technology is being used to locate potential sources of materials via online exchanges and resource maps. The materials and components in buildings can be fully catalogued using BIM so they can become materials banks for future generations. These inventories can be used to create a market for components within a building before it is demolished, which could be an incentive for it to be deconstructed rather than demolished, allowing items to be salvaged.

New business models show how thinking differently about ownership can enable valuable resources to be kept in high-value cycles, where manufacturers retain ownership of the product or incentivise return to ensure that components and materials are kept in circulation. There is also the potential for new local industries to be set up to remanufacture or reprocess products that would otherwise have been treated as waste.

Creating Revolutions

The Park 20|20 case study shows that it is possible to bring many of these design principles and concepts together to create buildings that are not only constructed from certified products and designed for adaptation and deconstruction, but also cost less to build, command higher rents and provide healthier internal environments.

The collaborative approach used by Delta Development allows it to have a direct relationship with the manufacturers and suppliers, which allows them to focus on materials and products selection, and to nurture innovative design solutions.

There is no need to wait for the policy landscape to shift or for the systemic problems in the industry to be resolved. With the right mindset, individuals and organisations can embrace this new agenda and create better buildings that leave a positive legacy for future generations.

References

INTRODUCTION

1. Ellen MacArthur Foundation, *Towards the Circular Economy: Economic and Business Rationale for an Accelerated Transition*, Cowes, UK: Ellen MacArthur Foundation, 2013.
2. OECD, 'Materials Resources, Productivity and the Environment', OECD Green Growth Studies, Paris: OECD Publishing, 2015, pp. 64, 84.
3. 'For many metals, this means that about three times as much material needs to be moved for the same quantity of metal extraction as a century ago.' 'The tendency to process lower grades of ore to meet increasing demand is leading to a higher energy requirement per kilogram of metals, and consequentially to increased production costs,' UNEP International Resource Panel, 'Decoupling 2: Technologies, opportunities and policy options', UNEP, 2014.
4. 'The extraction and use of material resources is closely linked to negative impacts on aspects of the environment,' UNEP, 'Decoupling 2'.
5. Ellen MacArthur Foundation, *Towards the Circular Economy*.
6. *Ibid*.

CHAPTER 1

1. Peiró, R. U., 'Material efficiency: rare and critical metals', *Philosophical Transactions*, The Royal Society, 2013.
2. Ellen MacArthur Foundation, *Towards the Circular Economy*.
3. House of Commons Environmental Audit Committee, 'Growing a circular economy: Ending the throwaway society, Third report of session 2014–15', London: The Stationery Office, 2014.
4. European Commission, 'Moving towards a circular economy', 2014, http://ec.europa.eu/environment/circular-economy/index_en.htm (retrieved 12 December 2014).
5. Ellen MacArthur Foundation, *Delivering the Circular Economy: A Toolkit for Policymakers*, Cowes, UK: Ellen MacArthur Foundation, 2015.
6. Stahel, W. R.: http://www.product-life.org/en/major-publications/the-product-life-factor, Product Life Institute, 1982 (retrieved 24 September 2014).
7. Braungart, M. and McDonough, W., *Cradle to Cradle: Remaking the Way We Make Things*, New York: North Point Press, 2002.
8. *Ibid.*, p. 72.
9. *Ibid.*, p. 76.
10. *Ibid.*, p. 75.
11. *Ibid.*, p. 61.
12. *Ibid.*, p. 105.
13. Pawlyn, M., *Biomimicry in Architecture*, London: RIBA Publishing, 2011.
14. Ellen MacArthur Foundation, *Towards the Circular Economy*.
15. CIBSE, 'TM56 Resource Efficiency of Building Services', London: CIBSE, 2014.
16. Recolight: www.recolight.co.uk, 2014 (retrieved 27 June 2014).
17. Clark, D. H., *What Colour is Your Building?* London: RIBA Publishing, 2013, p. 51 (a pie chart shows an example breakdown of construction embodied carbon in a new office building, with 13% and 42% associated with the substructure and superstructure, respectively).
18. Bioregional, 'Reclamation led approach to demolition', London: Defra, 2007.
19. World Economic Forum, 'Towards the circular economy: accelerating the scale-up across global supply chains', Geneva: World Economic Forum, 2014.

20 *Ibid*.

CHAPTER 2

1 Ellen MacArthur Foundation, *Towards the Circular Economy*.

2 CIBSE, 'Resource Efficiency of Building Services'.

3 BIS & Defra, 'Resource security action plan: making the most of valuable materials', London: Defra, 2012.

4 UNEP International Resource Panel, 'Decoupling 2'.

5 Circular Economy Task Force, 'Resource resilient UK', London: Green Alliance, 2013.

6 Allwood, J. and Cullen, J., *Sustainable Materials: With Both Eyes Open*, UIT Cambridge, 2012, p. 6.

7 Circular Economy Task Force, 'Resource resilient UK'.

8 House of Commons Environmental Audit Committee, 'Growing a circular economy'.

9 Circular Economy Task Force, 'Resource resilient UK'.

10 European Commission, 'Moving towards a circular economy', 2014, http://ec.europa.eu/environment/circular-economy/index_en.htm (retrieved 12 December 2014).

11 Prism Environment, 'Construction Sector Overview in the UK', EISC LTD, 2012, p. 9.

12 CIBSE, 'Resource efficiency of building services'.

13 Cited in: House of Commons Environmental Audit Committee, 'Growing a circular economy'.

14 *Ibid*.

15 All-Party Parliamentary Sustainable Resource Group, 'Remanufacturing towards a resource efficient economy', London: All-Party Parliamentary Sustainable Resource Group, 2014.

16 House of Commons Environmental Audit Committee, 'Growing a circular economy'.

17 Ellen MacArthur Foundation, *Towards the Circular Economy*.

18 Chatham House, 'A global redesign? Shaping the circular economy', Chatham House, London, 2014.

19 Ellen MacArthurFoundation, *Towards the Circular Economy*.

CHAPTER 3

1 O'Connor, J., 'Survey of actual service lives for North American Buildings', Canada: Forintek Canada Corp, 2004.

2 Habraken, N. J., *The Structure of the Ordinary: Form and Control in the Built Environment*, Cambridge, MA: The MIT Press, 2000, p. 6.

3 Flager, F. L., 'The design of building structures for improved life-cycle performance', Cambridge, MA: The MIT Press, 2003.

4 Edahiro, J., 'Rebuilding Every 20 Years Renders Sanctuaries Eternal – the Sengu Ceremony at Jingu Shrine in Ise', http://www.japanfs.org/en/news/archives/news_id034293.html, August 2013 (retrieved 5 February 2015).

5 McKeracher, C., 'Designing For Destruction: Anticipating Architectural Dismantling Through the Act of Making', Ottowa: Carleton University Ottawa, 2014.

6 Baum, A. M. and McElhinney, A., 'The causes and effects of depreciation in office buildings: A ten year update', 1996.

7 *Ibid*., p. 16.

8 Wilkinson, S. J., Remoy, H. and Langston, C., *Sustainable Building Adaptation: Innovations in Decision-Making*, London: RICS, 2014, Section 2.4.3.

CHAPTER 4

1 Button, H., Telephone interview between Howard Button and David Cheshire, 3 October 2014.

2 Addis, W. and Schouten, J., *Principles of Design for Deconstruction to Facilitate Reuse and Recycling*, London: CIRIA, 2014.

REFERENCES

CHAPTER 6

1 Brand, S., *How Buildings Learn – What Happens After They're Built*, London: Phoenix, 1994, p. 13.

CHAPTER 7

1 Aldersgate Group, *Resource efficient business models: The roadmap to resilience and prosperity*, London: Aldersgate Ltd, 2015.

2 Allwood, J. and Cullen, J., *Sustainable Materials*, p. 247.

3 Waste Resources Action Programme, 'Reducing and recycling plasterboard waste on a site where space is constrained', London: WRAP (n.d.).

4 Waste Resources Action Programme, 'Tate Modern 2, Case Study, Designing out Waste'. London: WRAP, 2010.

5 The Institute of Civil Engineers, 'ICE Demolition Protocol', ICE, 2008.

6 Bioregional, 'Reuse and recycling on the London 2012 Olympic Park: Lessons for demolition, construction and regeneration', London: Bioregional, 2011.

7 Cooper, T., *Longer Lasting Products: Alternatives to the Throwaway Society*, Farnham: Gower Publishing Limited, 2010.

8 CIBSE Journal, 'Trimming the fat', *CIBSE Journal*, 2014, pp. 34-35.

CHAPTER 8

1 Addis, W. and Schouten, J., *Principles of Design for Deconstruction*.

2 Habraken, J. N., *Supports, an alternative to mass housing*, London: The Architectural Press, 1972, translated by B. Valkenburg.

3 Working Commission W104 Open Building Implementation, http://www.open-building.org/ob/concepts.html (retrieved 11 September 2014) (n.d.).

4 Nascimento, D. M., *Vitruvius* (retrieved 12 September 2014, from Vitruvius: http://www.vitruvius.com.br/revistas/read/entrevista/13.052/4542?page=3).

5 Gassel, F. v., 'Experiences with the Design and Production of an Industrial, Flexible and Demountable (IFD) Building System', 2002.

6 Allwood, J. and Cullen, J., *Sustainable Materials*, p. 228.

7 DNV-Godstrup, 2005, www.dnv.rm.dk (retrieved 12 September 2014).

8 SEEDArchitects, 2010, Martini Hospital, http://www.seedarchitects.nl/page=site.home/lang=en#page-index(4) (retrieved 9 December 2014).

9 Brand, S., *How Buildings Learn*.

10 Ibid., p. 178.

11 Gazard, D., 'Abbey Mill, Bradford on Avon', Bradford on Avon Museum Booklet, 2012.

12 TEC Architecture and 3DReid, 'Adaptable design: Adapt and survive: A proposal for a form based design code', London: TEC Architecture and 3D Reid (n.d).

13 Telephone interview with Chris Gregory, TEC Architecture Limited, 2 March 2015.

CHAPTER 9

1 Addis, W and Schouten, J., *Principles of Design for Deconstruction*.

2 Ibid.

3 Arup and CIOB, 'Designing for the deconstruction process', Final report, Arup and CIOB, 2013.

4 Sassi, P., 'Closed Loop Material Cycle Construction', Cardiff: Cardiff University, School of Architecture, 2009.

5 BRE, *Dealing with Difficult Demolition Wastes: A Guide*, Watford: IHS BRE Press, 2013.

6 WellMet 2050, University of Cambridge, 'Novel Jointing Techniques to promote deconstruction of buildings', Cambridge: WellMet 2050, 2010.

7 Email correspondence with Jouke Post, 21 May 2015.

CHAPTER 10

1 Sassi, P., 'Closed Loop Material Cycle Construction'.

2 Ibid.

3 *Ibid.*

4 CIBSE, 'Resource Efficiency of Building Services'.

5 Braungart, M. and McDonough, W., *Cradle to Cradle*, p. 98.

6 Allwood, J. and Cullen, J., *Sustainable Materials*, p. 53.

7 WRAP, 'Building Insulation Foam Resource Efficiency Action Plan', London: WRAP, 2012.

8 RWM, *The Circular Economy: Exploring the Potential for your Business*, i2i, 2014.

9 Pawlyn, M., *Biomimicry in Architecture*. p. 35.

10 SGI Quarterly, 2013, www.sgiquarterly.org/feature2013jan-7.html (retrieved 4 March 2016).

11 *Ibid.*

12 Allwood, J. and Cullen, J., *Sustainable Materials*, p. 194.

13 WRAP, 'Guidance on re-use and recycling of used carpets and environmental considerations for specifying new carpet', London: WRAP, 2014.

14 *Ibid.*

15 Pearson, A., 'Fit for purpose', *RICS Modus*, February 2012, pp. 40-41.

16 Pawlyn, M., *Biomimicry in Architecture*, p. 39.

17 *Ibid.*

18 Natureplus e.v., 'Natureplus e.V Award Guideline GL0000 Basic Criteria', Natureplus, May 2011.

19 MBDC, 'Cradle to Cradle Certified Product Standard', Version 3.0, MBDC, 2013.

20 MBDC, LLC., 'Banned Lists of Chemicals, Cradle to Cradle Certified[CM] Product Standard, Version 3.0', McDonough Braungart Design Chemistry, LLC, 2012.

21 William McDonough + Partners, 'Sustainability Base: NASA's first space station on earth', San Francisco: William McDonough + Partners, 2012.

CHAPTER 11

1 Interview with Thomas Rau, 17 November 2014.

2 Maersk Line, 'Cradle to Cradle Passport – towards a new industry standard in ship building', OECD.org, http://www.oecd.org/sti/ind/48354596.pdf, 2011 (retrieved 19 June 2015).

3 UK-Green Building Council, 'UK-GBC Future Leaders: Innovation Project Summaries', UK-GBC, 2014.

4 Interview with Jan Jongert, Superuse Studios, 18 November 2014.

CHAPTER 12

1 Philips, 'Case study: RAU Architects', Koninklijke Philips Electronics N.V., 2012.

2 CIBSE, 'Resource Efficiency of Building Services'.

3 Ellen MacArthur Foundation, *Towards the Circular Economy*.

4 *Ibid.*

5 WRAP, 'Benefits of reuse case study: office furniture', London: WRAP, 2011.

6 WRAP estimate, private correspondence between WRAP and Rype Office, 2015.

7 Interview with Greg Lavery, Rype Office, 2 March 2015.

CHAPTER 13

1 Interview with Coert Zachariasse, 17 November 2014.

2 *Ibid.*

3 Manewa, A., Pasquire, C., Gibb, A. and Schmidt, R., 'A paradigm shift towards whole life analysis in adaptable buildings', Changing Roles: New Roles; New Challenges, Noordwijk aan Zee, 5–9 October, 2009.

Index

A

3D printing 91–4, 128–9
adaptability 22, 33, 53–63, 123
additive manufacture 91–4, 128–9
Architype 87–90
audits 47–8

B

benefits 15–17
biodegradable materials 84
biological materials 7, 8, 11, 65, 83–90
biomimicry 6, 91–2
brickwork 92–4
building in layers 33, 35–40, 55, 58, 62, 81–2
building life 19–22
building services 27, 39, 51, 62, 63, 86, 123
business models 33–4, 109–16

C

cardboard ductwork 86
cascading uses 8, 11, 83–4
certification systems 95–9, 118, 124
circular economy 3–5, 7–11, 31–4
ClickBrick 93–4
construction waste 46–7
Cradle to Cradle (C2C) 5–6, 95, 98–9, 118, 124

D

deconstruction 25–6, 65–79
demolition practices 25–30
Demolition Protocol 47–8
demolition wastes 29, 68–70
demountable buildings 55–6, 73–9
depreciation 22–3
design for adaptability 33, 53–63
design for disassembly 33, 65–79, 103, 123
design for reuse 65–79
design principles 33
designing-out waste 33, 41–52
disassembly 33, 65–79, 103, 123
documentation *see* certification systems; inventories

E

economic incentives 103–4, 113
economics 15, 124–5
end-of-life outcomes 7–10, 66
energy from waste 10, 69
energy performance 87, 111–12
environmental issues 9, 13–14
exchange forums 104–6, 129
extractive industries 14

F

finishes 6, 16, 37, 39, 47, 89
fixings 67–8, 83
flexibility 22, 44, 53, 55–6, 63
floor-to-ceiling heights 58–9, 62

I

incentives 103–4, 113
incineration 10
Industrial, Flexible and Demountable (IFD) 55–6
industrial symbiosis 6–7, 48
inventories 34, 104–5, 129

J

jointing techniques 22, 68, 92

L

layering approach 33, 35–40, 55, 58, 62, 81–2
lean design 50–1

135

M

markets for salvaged materials 103–4
masonry 92–4, 103
materials banks 102–3, 113, 129
materials exchange forums 104–6, 129
Materials Recovery Facilities (MRFs) 27–9
materials selection 33, 81–99
metals 8, 14, 27, 29, 90, 124
Multispace 58–9

N

NASA Sustainability Base 97–8
natural materials *see* biological materials
Natureplus 95
networks for salvaged materials 104–6, 129

O

obsolescence 22–3
Open Buildings 54–5

P

performance-based models 109–13
plasterboard alternative 84–6
plastics 29, 84, 90
PolyBricks 92–3
pre-demolition audits 47–8
procurement 118–19
product selection 33, 81–99

R

raw materials passports 102

reclaimed materials 47–50, 103–7
recycling 3, 8, 10, 26, 66, 90
refit and refurbishment 42–6
Refuse Derived Fuel 29
remanufacturing 9, 66, 113–15
reusing components and materials 9, 47–50, 65–79, 128

S

salvaged materials 47–50, 103–7
service-based models 8, 15–16, 103, 110–13, 115–16
standards 95–9
Suitebox 73–7
supply chain 117

T

technical materials 7, 8, 90–4
timber 8, 21–2, 29, 87, 94, 105–6
toxic materials 3, 6, 8, 65

V

value retention 10–11

W

waste as a resource 101–8
waste reduction 33, 46–7
waste to energy 10, 69
web platforms 105, 129
Werner Sobek 70–3

X

XX Architecten 77–9

Image credits

3D Reid	pp59, 61
AECOM – adapted from Manewa et al, A Paradigm Shift Towards Whole Life Analysis in Adaptable Buildings (2009)	p125
AECOM – based on the Multispace research by 3DReid	p58
Architype	p88
Jon Baker, Suitebox	pp74, 75, 76
Stewart Brand, Shearing Layers of Change (1994). Redrawn by AECOM	p36
David Cheshire, AECOM	pp26, 27, 28, 32, 54, 60
David Cheshire, AECOM – adapted from the DEGW 7 'S' model	p38
CIRIA	p66
Photographer: Matt Chisnall/courtesy of Derwent London	pp42, 43
CGI by Cityscape/courtesy of Derwent London	p44
Gert Jan den Daas, Daas Baksteen	p93
Reproduced by permission of DDG (Delta Developments Group) 2015	pp119, 120, 122
GatorDuct	p86
Givesha/courtesy Elina Grigoriou	p49
Chris Hill, Unity Works Board	p51
Richard Hind	pp20, 21
Matthias Koslik, Berlin	p71
Luuk Kramer	p79
Ellen MacArthur Foundation circular economy team drawing from Braungart & McDonough and Cradle to Cradle (C2C)	p7
Ellen MacArthur Foundation	p11
Make Architects	pp45, 46
National Union of Students	p112
Jouke Post	p78, 83

Thomas Robinson, Adaptavate	p85
Photographer: Cesar Rubio/courtesy William McDonough + Partners	pp96, 97, 98
Rype Office	p114
Sabin Design Lab., Cornell University – Jenny E. Sabin, Martin Miller, Nicholas Cassab	p92
Paola Sassi	p67
SEED Architects	pp56, 57
Slimline Buildings	p123
Walter R. Stahel, Product-Life Factor, a Mitchell Prize Winning Paper (1982)	p5
Werner Sobek	p72
Superuse Studios	p105, 106 ,107
Photographer: Sander van der Torren Fotografie, www.torren.nl. Reproduced by permission of DDG (Delta Developments Group) 2015	p121
UK Green Building Council. Redrawn by AECOM	p104